# Houghton Mifflin
## Primary
# Dictionary

**An American Heritage Dictionary**

**Houghton Mifflin Company** • Boston

Atlanta • Dallas • Geneva, Illinois • Palo Alto • Princeton, New Jersey

**Cover Photographs**

*Top left:* **Dalmatian,** Robert Pearcy, Animals, Animals;
*Top right:* **apple,** Fred Myers, Click/Chicago;
*Bottom left:* **heart,** Petroff Photography;
*Bottom right:* **telephone,** Sam Novak Photography, Ltd.

**Library of Congress Cataloging in Publication Data**

Krensky, Stephen.
  Houghton Mifflin primary dictionary.

  Rev. ed. of: My first dictionary.
  Summary: A dictionary for the primary grades explaining the meaning of words familiar to this age group and using each word in a sentence.
    1. English language – Dictionaries, Juvenile [1. English language – Dictionaries] I. Krensky, Stephen. II. Ulrich, George, ill. III. Houghton Mifflin Company. IV. Title
PE1628.5.K735    1986         423         85-27076
ISBN 0-395-38393-5

  Manufactured in the United States of America

# Contents

# Staff

**Editorial Director**
Fernando de Mello Vianna

**Project Editor**
Timothy J. Baehr

**Contributing Editor**
Vivian Fueyo
Lecturer, Department of Teacher Education
School of Education
California State University, Sacramento
Sacramento, California

**Definitions**
Stephen Krensky

**Art Director**
Geoffrey Hodgkinson

**Contributing Editor, Thesaurus**
Kerry W. Metz

**Illustrator**
George Ulrich

**Cover Design**
Ligature, Inc.
Chicago, Illinois

**Special Projects**
Kaethe Ellis
Reference Editor
Pamela B. DeVinne
Project Editor

**Proofreading**
Barbara Collins

**Consultants**

William K. Durr
Senior Author
Houghton Mifflin Reading Program
Professor of Education
Michigan State University
East Lansing, Michigan

Robert L. Hillerich
Professor of Education
Department of Educational
Curriculum and Instruction
Bowling Green State University
Bowling Green, Ohio

Hugh D. Schoephoerster
Reading and Language Arts Consultant
Houghton Mifflin Company
Boston, Massachusetts

John J. Pikulski
Professor of Education
Director, The Reading Center
University of Delaware

Bernice M. Christenson
Instructional Specialist, English/Reading
Los Angeles Unified School District
Los Angeles, California

Sarah N. Womble
Coordinator, Elementary Education
Pulaski County Special School District
Little Rock, Arkansas

Edmund H. Henderson
Director of Reading Education
University of Virginia

# Preface

*T*he *Houghton Mifflin Primary Dictionary* is intended for young children who are learning to read. Writers, educators, lexicographers, and artists collaborated to create a book that is both educationally sound and appealing to its young readers.

*The Houghton Mifflin Primary Dictionary* shares several features with dictionaries intended for older children and adults. Every word is defined, either directly by telling what it is or indirectly by building such a strong context that the meaning cannot escape understanding. The plural is given for each noun, and the past tense is given for each verb. Irregular forms are given as separate entries, and they are always cross-referred to the main entry containing the principal definition. Every word used in every definition is also an entry word. Illustrations are used to enhance definitions of things that can be pictured. They never appear merely for decoration or to accompany abstract concepts that cannot be pictured.

Great care was exercised in selecting the words to be included as entries. Children six or seven years old have an extensive spoken vocabulary. To include even half of it would require a dictionary of gigantic proportions or one with type too small for beginning readers. Therefore children's books and reading series were examined to determine which words would be most useful. Of the nearly 1,700 main-entry words in *The Houghton Mifflin Primary Dictionary,* 500 are those found most frequently in children's first primers and reading textbooks. The remainder of the vocabulary is a broad selection of other words children see and use every day.

There are over 600 full-color illustrations, averaging almost two per page. As mentioned before, each illustration has a definite defining purpose. But the pictures are also intended to invite the child to explore and browse.

A particularly attractive invitation to the dictionary appears at the very beginning. Four two-page spreads are devoted to instructional word games centered around four illustrated themes.

*The Houghton Mifflin Primary Dictionary* is a serious dictionary. It is also an enticing, inviting dictionary. We hope it will become one of the small treasures of children, who deserve not only to have command over their language but to love it as well.

# How to Use Your Dictionary

## What is a Dictionary?

A **dictionary** is a book about **words**. The dictionary tells you how to **spell** a word. The dictionary tells you **what a word means**. The dictionary shows you **how to use a word in a sentence**.

## How to Find a Word

The dictionary has words in **ABC order**. You have to know the **alphabet** to find a word in the dictionary. **Airplane** comes before **balloon**. **Balloon** comes before **cat**.

## ABCDEFGHIJKLMNOPQRSTUVWXYZ

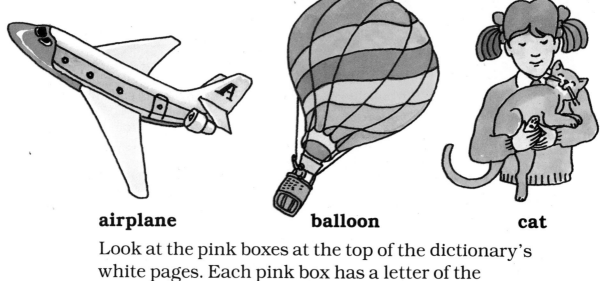

**airplane**          **balloon**          **cat**

Look at the pink boxes at the top of the dictionary's white pages. Each pink box has a letter of the alphabet in it. All the words that start with **A** are on the pages with ☐ A ☐. All the words that start with **B** are on the pages with ☐ B ☐. And all the words that start with **C, D**, **E**, **F** through **Z** are on the pages with ☐ C ☐ ☐ D ☐ ☐ E ☐ ☐ F ☐ and ☐ Z ☐ boxes.

# Looking Up a Word

Let's look at the letter **C**. All the words in the dictionary are in ABC order. And all the words that start with the letter **C** are in ABC order, too. Look at each letter of each word:

**cake** comes before **call**

**cake**
**call**

**came** comes before **camel**

**came**
**camel**

**can** comes before **candle**

**can**
**candle**

Let's look up **caterpillar**. Put your finger on each letter that spells **caterpillar**:

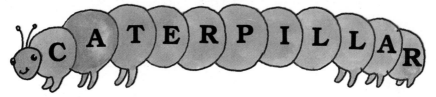

Look for the pink box with **C**. Find the beginning of the **C** words. Read the **C** words in ABC order until you find **caterpillar**. Where is **caterpillar**? **Caterpillar** comes in the ABCs after **catch** but before **cattle**:

**castle**

**cat**

**catch**

**caterpillar**

**cattle**

**Caterpillar** comes on **page** 48. The page numbers are at the bottom of the pages in blue numbers.

**Caterpillar** and the other words you look up are in **big, black letters.**

## Now Let's Use the Dictionary

1. Here is the alphabet. Some letters are missing. Say the letters that are missing.

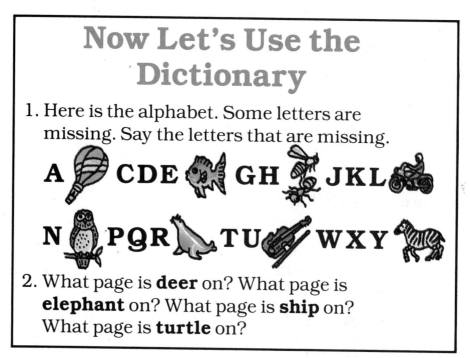

A CDE GH JKL

N PQR TU WXY

2. What page is **deer** on? What page is **elephant** on? What page is **ship** on? What page is **turtle** on?

## What You can Learn about a Word in the Dictionary

What does the dictionary tell you about **caterpillar**?

The dictionary *spells* the word.

The dictionary tells you the *meaning* of the word.

The dictionary often gives you a *picture* of the word.

**caterpillar**
A caterpillar is an insect. It looks like a worm covered with fur. **Caterpillars** change into butterflies or moths in cocoons.

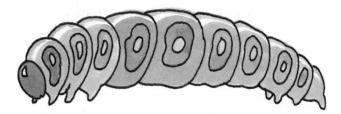

## Now Let's Use the Dictionary

1. Look up the words that the pictures show. What is the meaning of each word?

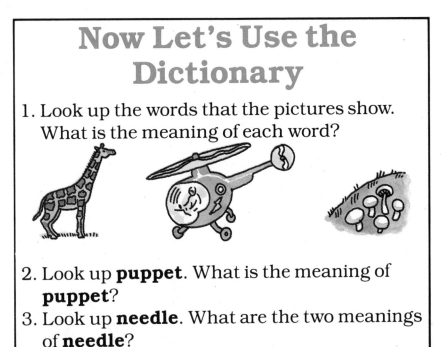

2. Look up **puppet**. What is the meaning of **puppet**?
3. Look up **needle**. What are the two meanings of **needle**?

## Different Words that are Spelled Alike

Some words are spelled alike but have different meanings. The dictionary puts **little black numerals** after these words. The numerals are a little above the big, black words:

### bat¹

A **bat** is a round piece of wood. One end of it is not as big as the other. **Bats** are used to hit baseballs.

### bat²

A **bat** is a small animal. It has a face like a mouse. **Bats** have wings covered with skin. They sleep upside-down during the day. They fly at night.

# Words about the Farm

## Let's Try This:

1. Look at ①. It is a **cow.**
   Read about the **cow** on page 63.
2. Look at ②. It is a **barn.**
   Read about the **barn** on page 19.
3. Look at ③. It is a **tractor.**
   Read about the **tractor** on page 299.
4. Look at ④. It is a **fence.**
   Read about the **fence** on page 96.
5. Look at ⑤. It is a **goat.**
   Read about the **goat** on page 116.
6. Look at ⑥. It is a **pumpkin.**
   Read about the **pumpkin** on page 224.

# Words about the City

MUSEUM OF ART

SHOE STORE

JONES STORE

## Let's Try This:

1. There are many places to visit in the city. Find the **library,** the **park,** the **museum,** the **hospital,** the **store,** and the **apartments.**

2. Where will you go first? Say **library, park, museum, hospital, apartments,** and **store** in ABC order.

SUBWAY

OAK APARTMENTS

LIBRARY

CITY HOSPITAL

AMBULANCE

CITY PARK

A-13

# Words about the Ocean

## Let's Try This:

1. Here is the **ocean.** What is another word for **ocean?** Find that word on page 246.

2. Look under the water. Find the **anchor.** Look up **anchor.** What is its meaning? Find the **fish,** the **lobster,** and the **whale.** What are the meanings of these words?

3. Look on top of the water. See the **boat,** the **ship,** and the **duck.** What do these words mean?

4. Look on the **beach.** Find the **dock,** the **waves,** and the **lighthouse.** What do these words mean?

# Words about Travel

## Let's Try This:

Here are pictures of things that go. Each thing has a number.

1. Put your finger on **①**. What is this object? Find it on page 46. Read the meaning.
2. Put your finger on **②**. What is this object? Find it on page 300. Read the meaning.
3. Put your finger on **③**. What is this object? Find it on page 303. Read the meaning.
4. Put your finger on **④**. What is this object? Find it on page 252. Read the meaning.

A-17

ABCDEFGHIJKLM
NOPQRSTUVWXYZ

Aa

## a

**1.** Red and blue are colors. Red is **a** color and blue is **a** color.

**2.** **A** means one. The river is one hundred miles long. It is also **a** hundred miles long.

## able

When you are **able** to do something, you can do it. David is **able** to build kites by himself.

## about

**1.** This book explains what words mean. This book is **about** words.

**2.** Billy and Jack can wear the same coat. They are **about** the same size.

## above

**Above** is the opposite of below. The wind blows the clouds across the sky. The clouds move far **above** the trees.

## accident

Two cars crashed into each other. This **accident** made a lot of noise. **Accidents** happen when people don't expect them.

## acorn

An **acorn** is a kind of nut. It grows on an oak tree. **Acorns** that are planted grow into oak trees.

## across

A large bridge was built from one side of a river to the other. The bridge was built **across** the river. Joan walked **across** the street to the store.

**acorn**

## act

**1.** In a school play Arthur pretends to be a king. Arthur **acts** in the play. He says the words a king should say. He does the things a king should do. He **acted** in the play four times.
**2.** Mary tries to be like her big sister. She **acts** like her big sister.

## actually

**Actually** means really. Barry thought he would not like summer camp. But once he got there, he **actually** enjoyed it.

## add

**1.** When you **add** two numbers, you find out how many they make together. Susan **added** 4 and 5. She got 9.
**2.** Mark puts flour into the bowl. Then he puts milk into the bowl. He **adds** the milk to the flour.

## address

An **address** is the name of a place. It tells where the place is.
Letters are sent through the mail with **addresses** on them.

**address**
On a mail box

**address**
On a letter

## afraid

**Afraid** means scared. Larry thought the lions at the zoo looked mean. He was **afraid** of them.

## after

**After** is the opposite of before. It is dark at night. It is light **after** the sun rises.

## afternoon

**Afternoon** is a part of a day. It comes between noon and sunset. **Afternoons** are short in the winter and long in the summer.

## again

When something happens more than once, it happens **again.** The snow fell for the first time last week. It fell **again** yesterday.

## against

1. Two teams play with each other. But only one team can win a game. The two teams play **against** each other to see which one will win.
2. Cindy held a blanket close to her face. She held it **against** her cheek.

## age

John is seven years old. His **age** is seven years. John and his friends are tall. They are tall for their **ages.**

## ago

Jack has played with his sled for three years. His parents gave it to him three years **ago.**

## agree

Tina and Kate share the same idea about horses. They **agree** that horses are nice animals. They **agreed** about this the first time they talked about it.

## air

When we breathe, we take **air** into our bodies. **Air** is all around us. But we cannot see it or touch it. We only feel **air** when it blows.

## airplane

An **airplane** is a machine. It has two wings and engines that make it fly through the air. **Airplanes** carry people from one place to another.

**airplane**

## alike

Most telephone poles look the same. Each one looks like the others. Telephone poles look **alike**.

**alike**

## alive

**Alive** is the opposite of dead. People, animals, and plants are **alive** while they live.

## all

There were three carrots in the refrigerator. Sharon ate the three carrots. She ate **all** the carrots in the refrigerator.

## alligator

An **alligator** is an animal. It is a reptile. It has a long body, short legs, and a long tail. **Alligators** have huge mouths. They live in swamps and rivers.

## almost

The leaves are red and yellow. Frost covers the ground. Winter will be here soon. It is **almost** here.

**alligator**

## alone

Sometimes Maggie likes to be by herself. She likes to be **alone** in her room.

## along

**1.** Joan walked on the grass next to the street. She walked **along** the street to the store.
**2.** Phillip brought a sandwich to school. He carried his lunch **along** with his books.

## aloud

Ann thought about a question during dinner. Then she asked the question **aloud.** Everybody heard it.

## alphabet

An **alphabet** is a group of letters. It is used to make a written language. Our **alphabet** is A, B, C, D, E, F, G, H, I, J, K, L, M, N, O, P, Q, R, S, T, U, V, W, X, Y, and Z. **Alphabets** are always written in a special order.

alphabet

## already

The sun rises at five o'clock. By this time, Jack is awake. He is **already** awake when the sun rises.

## also

Paul likes the beach. He likes to swim in the ocean too. Paul **also** likes to build castles in the sand.

## although

Nora wanted to stay awake for her birthday to begin. **Although** she was tired, she did not fall asleep.

## always

Spring comes after every winter. It never comes during any other part of the year. Spring **always** comes after winter.

## am

Jane will **be** seven years old next week. Then she will say "I **am** seven years old." **Am** is a form of **be.**

## among

Lisa likes many sports. She likes baseball and hockey better than most of the others. They are **among** her favorite sports.

## amount

A lot of snow is a large **amount** of snow. Large and small **amounts** of things can be measured.

## an

**An** is used instead of **a** with words that begin with A, E, I, O, or U. **An** elephant on **an** island that eats **an** apple or **an** orange is **an** unusual elephant.

## anchor

An **anchor** is a heavy piece of metal. **Anchors** are dropped into the water. They help boats and ships stay in one place.

anchor

## and

1. Bonnie walked to school with Jane. Bonnie **and** Jane walked to school together.
2. Oliver read a book. Then Nancy read it. Oliver read the book **and** then Nancy read it.
3. Six **and** two are eight.

### angry

Carl was mad at his sister. He was **angry**
with her because she had put a frog in his bed.

### animal

An **animal** is anything alive except plants.
People, dogs, cats, fish, birds, and insects are all
**animals.**

### another

There was a storm yesterday. There was a
second storm today. The first storm was
followed by **another** one.

### answer

"What color is the sky?" is a question. The
**answer** to this question is "Blue." Some hard
questions have long **answers.**

### ant

An **ant** is an insect. It is
black or red. **Ants** live
together in small
underground
tunnels.

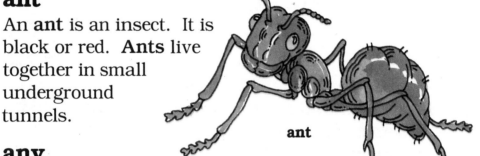

ant

### any

1. **Any** is another word
for **every.** Abby collects stamps. She keeps
**any** stamp she gets on a letter.
2. We ate all of the food we made for dinner.
There was not **any** food left after we ate.

### anybody

**Anybody** means any person. **Anybody** who
lives with other people does not live alone.

### anyone

**Anyone** means anybody. **Anyone** who lives
with other people does not live alone.

## anything

**Anything** means any thing. Babies chew on **anything** their teeth fit around.

## anytime

**Anytime** means at any time. **Anytime** you want to eat, let me know. I'll fix lunch **anytime** you're ready to eat.

## anyway

Michael went to sleep even though he was not tired. He went to sleep **anyway.** Michael had to get up early in the morning.

## apart

**1.** The trees are ten feet away from each other. They are ten feet **apart.**
**2.** Richard did not build his kite well. The wind pulled the paper away from the wood. The kite fell **apart** in the sky.

## apartment

An **apartment** is a place to live in a building. Some buildings have many **apartments** in them.

## apple

An **apple** is a kind of round fruit. It grows on a tree. **Apples** are red, yellow, or green.

## April

April is a month of the year. It has 30 days. **April** comes after March and before May.

**apple**

## aquarium

An **aquarium** is a glass box or bowl. It is filled with water. People who have fish or turtles at home keep them in **aquariums.**

## arch

An **arch** is a part of a building or bridge. It has a curve at the top. **Arches** hold up the part that is above them.

aquarium

arch

## are

Alex will go to the hockey game. Jim will **be** there too. Alex and Jim **are** on the hockey team. **Are** is a form of **be.**

## area

**1.** A field is part of a farm. It is an **area** of the farm. Fruits and vegetables can grow on different **areas** of land.
**2.** An **area** is a special place. The kitchen is the **area** where we cook and eat our food.

## aren't

**Aren't** is a short way to write **are not.** Clouds **aren't** green. They are white or gray.

## arm

An **arm** is a part of the body. It is between the shoulder and the hand. Amy used both **arms** to carry wood for the fireplace.

arm

## armor

**Armor** is a metal suit.  Knights wore **armor** to protect them when they fought.

armor

## army

An **army** is a group of people. These people fight together to protect their country. **Armies** from different countries fight each other on land.

## around

**1.** Ben built a fence on all sides of his yard.  He built a fence **around** his yard.
**2.** Vicki looked from one side of her room to the other.  She looked **around** her room for her shoe.

## arrive

Eric goes home after school. He gets there quickly.  He **arrives** at home a few minutes after school is over.  Yesterday he **arrived** late because he played baseball.

## arrow

**1.** An **arrow** is a thin piece of wood.  It has feathers at one end and a metal point at the other.  **Arrows** are shot from bows.
**2.** An **arrow** is a symbol.  It is used to point in one direction.

arrow

## art

Painting, poems, and music are all kinds of **art.**  These **arts** are thousands of years old.

## artist

An **artist** is a person who makes art. Some **artists** work on their art every day.

## as

Diane and Mary are the same height. Diane is **as** tall **as** Mary.

## ash

**Ash** is left when some things burn. It is soft and gray. The wind moved the **ashes** in the fireplace.

## ashamed

Sue felt bad because she lied to her sister. She was **ashamed** because she had said something that was not true.

## ask

**1.** The teacher says, "Who knows the answer?" She **asks** her class this question. She **asked** her class many questions.
**2.** Oliver says he wants another bowl of soup. He **asks** for the soup because he is hungry.

## asleep

In the middle of the night Roger is not awake. He is **asleep.** His eyes are closed. Roger rests when he is **asleep.**

## astronaut

An **astronaut** is a person whose job is to fly into space. **Astronauts** use rockets to get there. Some of them have walked on the moon.

**astronaut**

## at

**1.** Ellen went to the supermarket. She is there now. She is **at** the supermarket.
**2.** Nine o'clock is the time when school begins. School begins **at** nine o'clock.
**3.** Jeremy watched the bands in the parade. He looked **at** them as they passed him.

## ate

Chris **eats** three meals every day. He **ate** a big supper last night.
**Ate** is a form of **eat.**

## attention

The crowd watched a magician pull birds out of a handkerchief. Everybody watched the magician with care. The magician had the crowd's **attention.**

## attic

An **attic** is an area at the top of a house. People keep things in their **attics.**

attic

## August

**August** is a month of the year. It has 31 days. **August** comes after July and before September.

## aunt

Any sister of your mother or your father is your **aunt.** The wife of any of your uncles is also your **aunt.** Some people have several **aunts.**

## author

An **author** is a person who writes a story, a play, or a poem. Many **authors** write stories about themselves.

## automobile

**Automobile** is another name for **car.** It is a machine. It has four wheels, an engine, seats, and windows. There are always many **automobiles** on roads.

**automobile**

## autumn

**Autumn** is another name for **fall.** It is a season. It comes after summer and before winter. Leaves change color during the **autumn.**

## awake

When David is not asleep, he is **awake.** David is **awake** all of the time that he does not sleep.

## away

**1.** The ocean is three miles from here. It is three miles **away.**
**2.** Doug gave his jacket to a friend. He gave it **away** because it was now too small for him.
**3.** Ned walked from his room to the kitchen. He walked **away** from his room.

## awful

The cold wind blew the rain through the trees. The weather was terrible. It was an **awful** day.

## ax

**ax**

An **ax** is a tool. It has a sharp metal top and a long wood handle. The metal top is wide and flat. **Axes** are used to cut wood.

A B C D E F G H I J K L M
N O P Q R S T U V W X Y Z

Bb

## baby

A **baby** is a very young child.  **Babies** drink a lot
of milk.

## back

1. The **back** is a part of the body.  It is opposite
the chest and the stomach.  People ride
on the **backs** of horses.
2. **Back** is the opposite of front.  The engine is
in the front of the train.  The caboose is
at the **back.**
3. Students return to school in the fall.  They go
**back** to school in September.

## backward

**Backward** is the opposite of forward.
1. **Backward** spelled **backward** is ''drawkcab.''
2. Jack looked **backward** to see if a car was
behind him on the road.
3. Laurie fell on her back when she stepped
on the pillow on the floor.  Laurie fell
**backward.**

## bad

**Bad** is the opposite of good.
1. Cathy did not like the taste
of her sandwich.  It had a **bad** taste.
2. Amy is angry.  She shouts
at everybody.  Amy is in a **bad** mood.
3. Plants die in cold weather.  Cold
weather is **bad** for plants.

## bag

A **bag** is like a soft box.  It is used
to hold things.  It can be made
of paper, plastic, cloth, or leather.
Jason carried two **bags** filled
with food home from the store.

**bag**

## baggage

Penny takes two suitcases and a trunk to summer camp. She takes this **baggage** on the bus with her.

## bake

Jack cooks a pie in the oven. Jack **bakes** his pie for an hour. He **baked** two pies for a party.

## ball

A **ball** is a kind of toy.
It is round. Some **balls** are used in many sports.

baggage

ball

## balloon

A **balloon** is a closed bag filled with air. It is made of rubber or cloth. Some large **balloons** can lift people off the ground.

balloon

## banana

A **banana** is a kind of fruit. It has a curved shape and a yellow skin. Monkeys eat a lot of **bananas.**

## band

A **band** is a group of people who play music together. Many **bands** play music in parades.

banana

band

## bank

A **bank** is a safe place to keep money. Some **banks** are like a small box or jar. Other **banks** are huge buildings. People can leave money or borrow it at a **bank.**

## barber

A **barber** is a person whose job is to give haircuts. **Barbers** cut hair with scissors.

## bare

Ken did not have any socks or shoes on his feet. His feet were not covered by anything. Ken walked through his house on **bare** feet.

## bark

The **bark** is the skin of the tree. It covers the trunk and the branches.

## barn

A **barn** is a building. It is used on a farm. **Barns** are large and have few rooms inside them. Food for animals is kept in **barns.** Many farm animals stay in **barns** at night.

barn

## barrel

A **barrel** is used to hold things. It is made of wood or metal. The top and the bottom of **barrels** are flat circles. Some **barrels** have curved sides.

barrel

## base

1. A floor lamp stands on its bottom. The bottom is its **base**. Many objects stand on their **bases**.
2. A **base** is a heavy, square bag. It is used in a baseball game. After Rick hit the ball, he ran to first **base**.

## baseball

1. **Baseball** is a sport. It is played by two teams with a bat and a ball.
2. The ball used in a baseball game is a **baseball**. **Baseballs** are white.

baseball

## basket

A **basket** is used to hold things. It is usually made of wood or straw. Many **baskets** are shaped like bowls.

basket

## basketball

**1. Basketball** is a sport. It is played by two teams with a large ball and two baskets.
**2.** The ball used is called a **basketball. Basketballs** are usually orange.

## bat¹

A **bat** is a round piece of wood. One end of it is not as big as the other. **Bats** are used to hit baseballs.

## bat²

A **bat** is a small animal. It has a face like a mouse. **Bats** have wings covered with skin. They sleep upside-down during the day. They fly at night.

**basketball**

**bat²**

## bath

John sits in the water in the bathtub. He covers himself with soap. John takes a **bath** because he is dirty. He takes **baths** to get clean.

## bathroom

A **bathroom** is a room where you wash yourself. Hotels have many **bathrooms.**

## bathtub

A **bathtub** is a large hollow place where you take a bath. **Bathtubs** are filled with water.

### be

**1.** It is impossible for anyone to **be** in two places at the same time.

**2.** It would **be** strange to see an orange leaf in the spring.

**Am, is, are, was, were,** and **been** are forms of **be.**

### beach

A **beach** is an area of land. It is covered with sand. **Beaches** are at the edge of lakes or oceans.

### beak

A **beak** is part of the mouth of a bird. It sticks out from the bird's face. **Beaks** are hard and pointed.

### bear

A **bear** is a large animal. It has thick fur and long claws. **Bears** sleep in caves during the winter.

**beak**

**bear**

## beat

**1.** Fred mixes the eggs and milk with a fork. He **beats** the eggs and milk. He **beat** them for three minutes.

**2.** Richard hits his drum in the parade. He **beats** the drum as part of the music.

**3.** Our team won all of its soccer games this fall. We **beat** all of the other teams in town.

**4.** Bill could feel his heart move in his chest. He feels the **beat** of his heart after he runs home from school.

## beaten

**1.** Fred beats the eggs and milk. He has **beaten** them for three minutes.

**2.** Richard has **beaten** the drum during the parade for 15 minutes.

**3.** Our soccer team has **beaten** all the other teams in town.

**Beaten** is a form of **beat**.

## beautiful

**Beautiful** means very pretty. Rainbows are **beautiful**.

## beaver

A **beaver** is an animal. It has a big, flat tail and large front teeth. **Beavers** can chew through trees. They build dams in rivers.

beaver

## became

A caterpillar **becomes** a moth inside a cocoon. The caterpillars **became** moths during the summer.

**Became** is a form of **become**.

## because

Elizabeth was hungry. So, she made a sandwich and ate it. She ate the sandwich **because** she was hungry.

## become

A caterpillar changes into a moth. It **becomes** a moth inside a cocoon. The caterpillars **became** moths during the summer.

## bed

A **bed** is a kind of furniture. People sleep on **beds.**

## bee

A **bee** is an insect. It can fly. **Bees** make honey.

## beef

Beef is a kind of meat. It comes from cattle. Hamburger is made from **beef.**

bee

## been

**1.** It is impossible for anyone to **be** in two places at the same time. Nobody has **been** in two places at the same time.
**2.** It would have **been** strange to see an orange leaf in the spring.
**Been** is a form of **be.**

## before

**1. Before** is the opposite of after. Every day the sun rises in the morning and sets in the evening. Every day the sun rises **before** it sets.
**2.** Alice rode on an airplane for the first time last week. She had never been on a plane **before.**

## began

The alphabet **begins** with the letter "A." This dictionary **began** with the word "a."
**Began** is a form of **begin.**

## begin

**Begin** is the opposite of end. "A" is the first letter in the alphabet. The alphabet **begins** with "A." This dictionary **began** with the word "a."

## begun

This dictionary **begins** with the word "a." Dictionaries have always **begun** with the word "a." **Begun** is a form of **begin.**

## behave

If the twins are good, John will read to them. If they **behave,** he will read their favorite story. They **behaved** yesterday.

## behind

Sally and Julie stood in line at the movie theater. Sally stood in front of Julie. Julie stood **behind** Sally.

## believe

Patrick thinks it is true that dragons live in the mountains. He **believes** that they live there in caves. Patrick first **believed** this after he saw smoke rise above the mountains.

## bell

A **bell** is a hollow piece of metal. It is shaped like a cup that is upside-down. **Bells** ring when they are hit. Some of them are used to make music.

**bell**

## belong

**1.** A fish should stay in the water. It **belongs** in the water.

**2.** These shirts **belong** to George. They are his shirts.

## below

**Below** is the opposite of above. Roots grow under the surface of the earth. Roots grow **below** the surface.

## bend

The blacksmith changes the shape of a piece of metal. He puts a curve in it. He **bends** the metal to make a horseshoe. He **bent** four pieces of metal into horseshoes for one horse.

## bent

The blacksmith **bends** a piece of metal to make a horseshoe. He **bent** four pieces of metal into horseshoes for one horse.
**Bent** is a form of **bend.**

## berry

A **berry** is a kind of fruit. It is round and small. **Berries** have many colors and grow on bushes in large groups. **Berries** don't grow during the winter.

**berry**

## beside

Louise sat next to Jennifer on the bus. They sat **beside** each other all the way to school.

## best

**Best** is the opposite of worst. Susan's **best** friend is the friend she likes more than anyone else.

## better

**Better** is the opposite of worse. Mark does not swim as well as Edward. Edward swims **better** than Mark.

## between

**1.** C-A-T spells cat. The A is **between** the C and the T.
**2.** Janet cannot decide which birthday present to open first. She is curious about them both. But she must choose **between** them.
**3.** Two baseball teams play each other today. There is a game **between** them today.

## beyond

Jackie walks over the field and into the forest. She walks **beyond** the field.

## bicycle

A **bicycle** is a machine. It has two big wheels. It does not have a motor. People use their legs to make **bicycles** go.

**bicycle**

## big

**Big** is the opposite of small.

**1.** Big means large. Bears are **big.** Goldfish are not.

**2.** One thousand apples is a lot of apples. One thousand is a **big** number.

## bill

A **bill** is the beak of a bird. Parrots have big **bills.**

## bird

A **bird** is an animal. It is covered with feathers. It has two wings. Robins, chickens, eagles, and ostriches are all **birds.** Most **birds** can fly.

bill

## birth

The **birth** of a baby happens the moment that it is born. **Births** usually happen in hospitals. Many babies were born yesterday.

bird

## birthday

Lynne was born on November 28. November 28 is her **birthday.** People have **birthdays** each year on the day they were born.

## bit

The hungry dog **bites** into a piece of meat. The dog **bit** into the meat because the dog was hungry.

**Bit** is a form of **bite.**

## bite

**1.** The dog puts its teeth into a piece of meat. It **bites** into the meat. The dog **bit** into the meat because it was hungry.
**2.** Alice took a **bite** out of a sandwich. She chewed it slowly. After she took ten **bites,** the sandwich had disappeared.

## bitten

The dog **bites** into a piece of meat. The dog had **bitten** into the meat because it was hungry.
**Bitten** is a form of **bite.**

## bitter

**Bitter** foods have a bad taste.

## black

**Black** is a color. It is the opposite of white. Most books are printed with **black** ink.

## blacksmith

A **blacksmith** is a person whose job is to make things out of iron.
**Blacksmiths** make horseshoes.

**blacksmith**

## blanket

A **blanket** is a large piece of soft cloth. It can be thick or thin. People use **blankets** in bed to keep warm.

**blanket**

## blew

1. The wind **blows** the dead leaves from the branches. It **blew** them toward the ground.
2. Jeremy **blew** the trumpet in the band.
3. The trumpets **blew** in the parade.
**Blew** is a form of **blow.**

## blind

People see with their eyes. Anybody who cannot see is **blind.**

## block

1. A **block** is an object shaped like a rectangle or a square. It can be made of many things. Babies play with **blocks** of wood.
2. A **block** is an area of a city. It has four sides. The sides of a city **block** are streets.

**block**

## blood

**Blood** is a red liquid. It goes through all people and many animals. It carries important foods and other things. Nobody can live without **blood.**

## blossom

A **blossom** is a flower on a bush or a tree. The **blossoms** on apple trees are white or pink.

## blow

1. The wind **blows** the dead leaves from the branches. It **blew** them to the ground.
2. Jeremy makes sound come out of the trumpet. He **blows** the trumpet in the band.
3. Loud sounds come out of trumpets. Trumpets **blow**.

**blossom**

## blown

1. The wind **blows** the dead leaves from the branches. It has **blown** many others off the tree already.
2. Jeremy has **blown** the trumpet for two years.
3. The trumpets have **blown**.
**Blown** is a form of **blow.**

## blue

**Blue** is a color. The sky is **blue** when no clouds cover it.

## board

A **board** is a piece of wood. It is shaped like a rectangle. Carpenters use many **boards** to build houses.

**board**

# boat

A **boat** carries people and things over the water. It is shaped like a bowl with a point on one end. **Boats** are made of wood, metal, or other things. Many **boats** have engines to make them move. Other **boats** have sails. Some **boats** are big enough to live on.

boat

# body

The leg of a person is part of a person. The **body** of a person is the whole person. All people and animals have **bodies.**

# boil

When water gets very hot, it **boils.** Bubbles form and steam rises into the air. The water in the pot **boiled** for the soup.

# bone

A **bone** is a part of the body. It is hard. **Bones** go through the body. We can feel the **bones** in our fingers and elbows.

# book

A **book** is a group of pages. They are covered with words printed by a machine. The pages are held together with thread or glue. This dictionary is a **book.** Many **books** also have pictures.

## boot

A **boot** is a tall shoe. It fits over the foot and part of the leg. Most people wear **boots** in the snow.

boot

## border

A **border** is an edge. The **border** of the lake is the shore. **Borders** are lines that divide one place from another.

## born

Susan is one day old. She was **born** yesterday.

## borrow

Jane takes books from the library. She returns them a few days later. She **borrows** the books to read. Last week she **borrowed** five of them.

## both

Gary has two hands. He uses them to put on a tight pair of boots. He needs **both** hands to put on **both** boots.

## bother

When something gives Jay trouble, it **bothers** him. Jay's little sister **bothered** him with a lot of questions.

## bottle

A **bottle** is used to hold liquids. It is made of glass or plastic. **Bottles** are not as big at the top as they are at the bottom. Some **bottles** are easy to break.

bottle

## bottom

**Bottom** is the opposite of top. Our heads are at the top of our bodies. Our feet are at the **bottom**. When Rachel walks in bare feet, the **bottoms** of her feet get dirty.

## bought

David **buys** food with money. He **bought** two bags of food at the supermarket. **Bought** is a form of **buy**.

## bow[1]

**1.** A **bow** is a curved piece of wood. A piece of string is tied to both ends of it. **Bows** are used to shoot arrows.

**2.** A **bow** is made of circles of ribbon or string. Joan tied a **bow** on the top of a present for her father.

**bow**[1]
To shoot arrows

**bow**[1]
Circles of ribbon

## bow[2]

The knight bends forward. He **bows** to the king. The knight **bowed** to the king with respect.

## bowl

A **bowl** is used to hold things. It is shaped like half of a ball. **Bowls** are hollow in the middle. People eat soup and cereal out of them.

**bowl**

## bowling

**Bowling** is a sport. Any number of people can play it. In **bowling,** a heavy ball is used to knock over pieces of wood shaped like bottles.

## box

A **box** is made to hold things. All **boxes** have four sides and a bottom. Some **boxes** have tops.

**bowling**

## boy

A **boy** is a male child. **Boys** grow up to be men.

## brain

The **brain** is a part of the body. It is inside the head. People and animals have **brains. Brains** make arms, legs, eyes, and ears able to work together. People think with their **brains.**

## branch

A **branch** is a part of a tree. **Branches** stick out from the trunk of a tree.

## brave

Eleanor was afraid to go outside at night. But one time she wanted to look at the stars. She was **brave** enough to go outside even though she was scared.

**box**

## bread

**Bread** is a kind of food. It is made of flour, milk, and other things mixed together. **Bread** is baked in an oven.

## break

**1.** A whole window is in one piece. If it **breaks,** it becomes several small pieces of glass. One window **broke** when a baseball went through it.
**2.** When a machine **breaks,** it will not work. It must be fixed before anybody can use it.
**3.** People who rob banks do something against the law. They **break** the law.

## breakfast

**Breakfast** is a meal. It is the first meal of the day. Cereal, eggs, pancakes, juice, and milk are all parts of **breakfasts.**

## breath

We take air into our bodies through our mouths. Each amount of air we take is a **breath.** Everyone takes many **breaths** in a minute.

## breathe

People and many animals take in and push out air. They **breathe** air through their mouths. Jeff **breathed** very fast after he ran home from school.

## bridge

A **bridge** joins two different places. **Bridges** are built over rivers so that people can cross the rivers without a boat.

**bridge**

# bright

**1.** The sun shines with a lot of light. The sun is **bright**.

**2.** Jack wears an orange shirt when he rides his bicycle at night. His shirt is so **bright** that everyone can see him.

# bring

Carol takes a present to Nora's birthday party. She **brings** the present for Nora. Carol **brought** the present into Nora's room.

# broke

**1.** If a window **breaks,** it becomes several small pieces of glass. One window **broke** when a baseball went through it.

**2.** After the machine **broke,** it would not work. The machine must be fixed before anybody can use it.

**3.** The people who robbed the bank **broke** the law.

**Broke** is a form of **break.**

# broken

**1.** If a window **breaks,** it becomes several pieces of glass. This window has **broken** into five pieces.

**2.** The machine has **broken.** It will not work.

**3.** People who rob banks have **broken** the law.

**Broken** is a form of **break.**

# broom

A **broom** is a kind of tool. It is made of pieces of straw tied together to a long wood handle. Most people use **brooms** to clean dirt off floors.

**broom**

### brother

A boy is a **brother** to the other children in his family. Some families have several **brothers** and sisters in them.

### brought

Carol **brings** a present to Nora's birthday party. She **brought** the present into Nora's room. **Brought** is a form of **bring**.

### brown

**Brown** is a color. Chocolate is **brown**.

### brush

**1.** A **brush** looks like a small broom. Some **brushes** are used to clean things. Some **brushes** are used to paint with.

**brush**
For paint

**brush**
For the teeth

**2.** Every morning Alan cleans his teeth. He **brushes** them carefully. He **brushed** his teeth every morning this week.

### bubble

A **bubble** looks like a balloon. It is round and has air in it. Some **bubbles** are made of soap and water. Some **bubbles** form when water boils.

### bucket

A **bucket** is used to hold things. It is made of wood, metal, or plastic. **Buckets** have handles. They are round at the top and the bottom.

**bucket**

## build

A carpenter puts houses together. He **builds** houses. Three carpenters **built** a house in one summer.

## building

A **building** is a place where people live, work, or play. Houses, hotels, schools, and barns are all **buildings.**

## built

A carpenter **builds** houses. Three carpenters **built** a house in one summer. **Built** is a form of **build.**

## bull

A **bull** is a large animal. It is a kind of cattle. It has long horns. **Bulls** like to eat grass.

bull

## bulldozer

A **bulldozer** is a very large tractor. It pushes dirt and rocks from one place to another. **Bulldozers** help shape the ground where buildings will be built.

bulldozer

## bump

**1.** Duncan hits his head on the attic ceiling. He **bumps** his head because the ceiling is low. Yesterday he **bumped** his head twice in the attic.
**2.** A hill is a giant **bump** on flat ground. **Bumps** are round at the top.

## burn

The logs in the fireplace are covered with fire. They **burn** for a long time. The logs **burned** for hours in the fireplace.

## bus

A **bus** is a machine. It has four wheels, an engine, and many seats and windows. It can carry many people. School **buses** are yellow.

bus

## bush

A **bush** is a plant. It has many wood branches and green leaves. **Bushes** are not as big as trees. Some fruits, flowers, and vegetables grow on **bushes**.

## busy

Lynne has a lot of work to do. Lynne is very **busy** with her work. People who are **busy** with their work cannot take time to play games.

## but

**1.** Some bears look friendly. Few bears are friendly. Bears look friendly, **but** people should stay away from them.
**2.** Almost everybody liked the new car. Albert did not like it. Everybody **but** Albert liked the new car.

## butter

**Butter** is a kind of food. It is soft and yellow. We cook with **butter.** Some people put it on bread. **Butter** is made from the milk and cream that comes out of a cow.

## butterfly

A **butterfly** is an insect. It has four thin wings. These wings have bright colors. Caterpillars change into **butterflies.**

**butterfly**

## button

A **button** is a round object. It is about the size of a coin. **Buttons** are sewed on clothes. They help keep shirts and jackets closed.

**button**

## buy

David spends his money at the supermarket. He **buys** food with the money. David **bought** two bags of food there.

## by

**1.** Dana asked her teacher a question. The question was asked **by** Dana.
**2.** Penny should not be awake after nine o'clock. She should be asleep **by** then.

A B C D E F G H I J K L M
N O P Q R S T U V W X Y Z

Cc

## caboose

A **caboose** is a kind of railroad car. It is the last car on a train. Few trains now have **cabooses.**

caboose

## cage

A **cage** is a metal or wood box. It has many open spaces in its sides. Some animals at a zoo are kept in **cages.**

cage

## cake

A **cake** is a dessert. It is made of flour, sugar, eggs, and milk. **Cakes** are baked in ovens before anyone eats them.

43

## call

**1.** Diane speaks to her friends on the telephone. She **calls** them at home. She **called** three of them yesterday.
**2.** "Where are you, George?" shouts Bonnie. She calls for him all over the house.
**3.** Anne has a doll. The doll's name is Emily. Anne **calls** the doll Emily.
**4.** Nina talked to Julie on the telephone. Nina made the **call** from the kitchen. She made some other **calls** from her room.
**5.** The police heard a shout from inside a bank. They heard a **call** for help.

## calm

**1.** Almost nothing ever bothers Jason. He is usually very **calm,** even when things are exciting.
**2.** The wind is not blowing. The air is **calm.**

## came

**1.** Bill and Sue **come** toward the kitchen. They **came** to the kitchen to eat dinner.
**2.** The circus **came** to town last month.
**3.** The apple juice only **came** in bottles.
**Came** is a form of **come.**

## camel

A **camel** is a large animal. It has long legs and a long neck. It also has one or two humps on its back. **Camels** do not need to drink water very often. People ride them across deserts.

**camel**

## camera

A **camera** is a small machine. It makes pictures or movies. Some **cameras** make finished pictures in seconds.

camera

## camp

1. Sally and her family take trips through the forest. They sleep in tents and cook with fires. They **camp** in the forest every summer. Last year they **camped** near the lake.
2. Many children go to a **camp** in the summer. They learn about nature and play games there. Some **camps** play games against one another.

## can[1]

A **can** is a metal object. It is shaped like a small barrel. Some foods are sold in **cans.**

## can[2]

Tom knows how to whistle. He **can** whistle all of his favorite songs. He **could** whistle a short song with one breath.

## candle

A **candle** is a stick of wax. String goes through the middle of it from top to bottom. **Candles** are burned to make light.

candle

## candy

**Candy** is a kind of sweet food. **Candies** may have chocolate, nuts, or fruit.

## cannot

**Cannot** is the opposite of can. A dog can run, but it **cannot** fly.

## canoe

A **canoe** is a narrow boat. It can only carry a few people. **Canoes** can travel quickly.

### can't

**Can't** is a short way to write **cannot**. Fish can swim underwater, but they **can't** talk.

### cap

A **cap** is a small hat. People who play baseball wear **caps**.

### car

**1.** A **car** is a machine. It has four wheels, an engine, seats, and windows. People travel over roads in **cars**.

**2.** A **car** is one part of a train. It is like a big box on wheels. It can carry people, animals, machines, foods, or other things.

cap

car

### card

**1.** A **card** is a piece of thick paper. It is shaped like a rectangle. **Cards** have numbers and symbols on them. People play games with **cards**.

**2.** Joyce got a funny birthday message from her brother. He sent her a funny **card**.

card

## care

**1.** Jenny played with her baby brother while their parents went to a movie. She also fed him and read him a story. Jenny took **care** of her brother while their parents were away.
**2.** It is easy to break kites. They must be handled with **care** or they will break.
**3.** Nancy's dog is sick. She is worried about her dog. Nancy **cares** about her dog. She **cared** about her cat when it was sick too.

## careful

Jan draws a flower with care. She is very neat. Jan is **careful** to make the picture look like a flower.

## carpenter

A **carpenter** is a person whose job is to build things out of wood. Several **carpenters** build a house together.

## carrot

A **carrot** is a kind of vegetable. It is long and orange. **Carrots** grow underground.

## carry

Cindy and Meg take their lunches to school. They **carry** their lunches in brown bags. Everyone in their class **carried** a lunch to the picnic.

carrot

## cartoon

A **cartoon** is a kind of picture. Artists draw the people and animals in **cartoons** in a simple way. Some **cartoons** are made into movies.

## castle

A **castle** is a large building. It is made of stone. **Castles** were built hundreds of years ago as safe places for people to live. Kings, queens, and knights lived in **castles.**

castle

## cat

A **cat** is a small animal. It has four legs, soft fur, and a long tail. Many people keep **cats** as pets. Most **cats** and dogs do not like each other.

cat

## catch

Many balls are hit into the air in a baseball game. People in the game **catch** some of the baseballs before they reach the ground. They **caught** the baseballs in their gloves.

## caterpillar

A **caterpillar** is an insect. It looks like a worm covered with fur. **Caterpillars** change into butterflies or moths.

caterpillar

## cattle

**Cattle** are animals with horns. They are raised for milk and meat. Cows and bulls are **cattle**.

## caught

People in a baseball game **catch** some of the baseballs hit into the air. They **caught** the baseballs in their gloves.

**Caught** is a form of **catch**.

cattle

## cause

**1.** Lightning can make forest fires happen. Lightning **causes** trees that it hits to burn. Lightning **caused** a barn to burn during a bad storm.

**2.** The forest fire began during the storm. The **cause** of the forest fire was lightning. There are many different **causes** of forest fires.

## cave

A **cave** is a hollow area in the ground. It is only open on one side. Some bears sleep in **caves** in the winter.

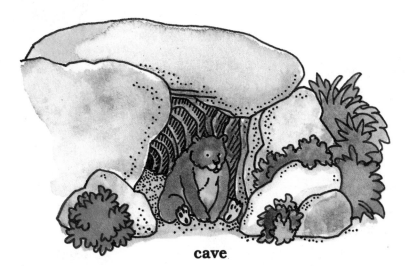

cave

## ceiling

A **ceiling** is the top side of a room. Martha walks on the floor. She looks up at the **ceiling.** Some **ceilings** have lamps that hang from them.

## cellar

A **cellar** is a space at the bottom of a house. Many **cellars** are built underground.

## cent

A **cent** is a coin. One penny is one **cent.** A nickel is the same amount of money as five **cents.**

## center

The **center** of an object is the middle of it. The **centers** of many doughnuts are holes.

## cereal

**Cereal** is a kind of food. Most people eat **cereal** with hot or cold milk in a bowl. Many **cereals** are made from corn, wheat, or rice.

## chair

A **chair** is a kind of furniture. It is made of wood, metal, or other things. It has four legs and a seat. People sit on **chairs.**

## chance

**1.** The sky is gray today. It may rain. There is a **chance** it will rain. As the sky gets dark, the **chances** get better that it will rain.

**2.** Anything that happens by luck or accident happens by **chance.**

**chair**

## change

**1.** Leaves are green in the summer. They become red, orange, or yellow in the fall. Leaves **change** color in the fall. Last year most of the leaves **changed** color in October.
**2.** Arthur takes off one shirt and puts on another. He **changes** his shirt because it got wet in the rain.
**3.** Any number of coins is an amount of **change**. How much **change** does Joyce have in her pocket?

## chase

**1.** Our dog runs after cars. It **chases** cars that pass on our street. Yesterday our dog **chased** a truck.
**2.** Tom watched one car follow another in a movie. Both cars went very fast. The **chase** went through a city. Many car **chases** end in accidents.

## check

**1.** Mary makes sure her dog is inside before she goes to sleep. She **checks** to see where her dog is. One night she could not find her dog until she **checked** under her bed.
**2.** Leslie's shirt had little squares all over it. Each square was a **check.** Her shirt was covered with red and white **checks.**
**3.** The teacher made a mark next to the answers on a test. He put a **check** next to the right answers. This is a **check** (✔).

**check**

## cheek

A **cheek** is a part of the face. It is below the eye and next to the nose. Every face has two **cheeks.**

## cheerful

**Cheerful** people are happy people who share how they feel with others.

## cheese

**Cheese** is a kind of food. It is made from milk. Many **cheeses** are yellow.

## cherry

A **cherry** is a kind of fruit. It is small, round, and red. **Cherries** grow on trees.

**cherry**

## chest

**1.** The **chest** is a part of the body. It is below the shoulders and above the stomach. When people breathe, their **chests** move.
**2.** A **chest** is a strong box. It is made of wood or metal.

**chest**

## chew

Edward eats a sandwich for lunch. He **chews** each bite of it with his teeth. Then he swallows it. He **chewed** one bite ten times.

## chicken

1. A **chicken** is a kind of bird.
It cannot fly very far. **Chickens** grow
up to be hens or roosters.
2. **Chicken** is a kind of meat.
It comes from a **chicken.**

## chief

Many firefighters help put out
a fire. Their **chief** tells them
what to do. **Chiefs** are people
who tell other people what to do
in their jobs.

## child

A **child** is a boy or a girl. **Children**
grow up to be men and women.

## children

**Children** means more than one **child.**
**Children** grow up to be men and women.

## chimney

A **chimney** is a part of a house.
The smoke from a fireplace goes
up the **chimney.** The tops
of tall **chimneys** stick out
of roofs.

## chin

The **chin** is the bottom
of the face. **Chins** move
when people talk.

## chocolate

1. **Chocolate** is a kind
of candy. It is usually brown.
2. **Chocolate** is a flavor. Joyce likes
**chocolate** birthday cake better than any other
kind.

**chicken**

**chimney**

53

### choose

Abby can only have one balloon. She must **choose** just one. Abby **chose** the red one.

### chose

Abby must **choose** just one balloon. She **chose** the red one.
**Chose** is a form of **choose.**

### chosen

Abby must **choose** a balloon. She has **chosen** the red one.
**Chosen** is a form of **choose.**

## Christmas

**Christmas** is a holiday. It comes each year on December 25.

### circle

A **circle** is a round, flat shape.
**Circles** have no corners.

circle

### circus

A **circus** is a kind of show. It has many people and animals in it. They do many different tricks for the crowd. Clowns make the crowd laugh at **circuses.**

circus

## city

A **city** is a large town. It has many streets and tall buildings. A lot of people live in **cities.**

## class

A **class** is a group of students who learn together at school. There are **classes** in every grade.

## claw

**1.** A **claw** is part of an animal's foot. It has a sharp, curved nail on the toe. Birds hold on to branches with their **claws.**

**2.** A lobster **claw** is also a part of the lobster.

**claw**
Of a bird

**claw**
Of a lobster

## clay

**Clay** is a kind of earth. It is used to make pots, cups, and other objects.

## clean

**1. Clean** is the opposite of dirty. Walter washed his hands with soap. His hands were **clean** after he washed them.

**2.** Walter gets the dirt off his hands. He **cleans** them with soap and water. After he did that, he **cleaned** his face.

## clear

It was easy to see through the window. The window had no color. It was **clear.**

## clever

Jamie can think quickly and he has many good ideas. Jamie is **clever.**

## climb

The monkey uses its arms and legs to go up a tree. It **climbs** up the tree. It **climbed** down later.

## clock

We look at a **clock** to see what time it is. Some **clocks** have bells in them that ring every hour.

**clock**

## close

**1.** Flowers grow near the stone wall in the garden. They grow **close** to the wall.
**2. Close** is the opposite of open. People buy food at the store from nine o'clock in the morning until eight o'clock at night. The store **closes** at eight o'clock. It was **closed** all day on Christmas.
**3. Close** is the opposite of open. Fred shuts the book. He **closes** the book because he has finished it.

## closet

A **closet** is a very small room. It is a place where clothes, shoes, and other things are kept. Most houses have several **closets.**

## cloth

**Cloth** is used to make clothes. It is made from plants, animal fur, and other things. Cotton and wool are made into many different kinds of **cloths.**

## clothes

Coats, shirts, dresses, pants, and socks are all kinds of **clothes.** People wear thick **clothes** in the winter to keep warm.

## cloud

A **cloud** is made of many tiny drops of water. It can be white or gray. The wind blows the **clouds** across the sky.

## clown

A **clown** is a person whose job is to make people laugh. **Clowns** wear paint on their faces. They play tricks on one another. Many **clowns** work in a circus.

**clown**

## coat

A **coat** is a long and heavy jacket. Wool **coats** are warm.

## cocoon

A **cocoon** is a ball of threads. Some caterpillars make them. They live in **cocoons** for a while as they change into moths.

## coin

A **coin** is a kind of money. It is made of metal. Most **coins** are round. Pennies, nickels, dimes, and quarters are all **coins.**

**cocoon**

## cold

**1. Cold** is the opposite of hot. Snow and ice are **cold.** Fires are not.

**2.** The **cold** of winter turned the water in the pond to ice.

**3.** Cathy feels a little sick. She sneezes a lot. She is very tired. Cathy has a **cold.** She gets **colds** several times during the winter.

## collect

Eliza has all of the books by her favorite author. She **collects** them. She **collected** every one of the books that she could find.

## color

Everything has a **color.** We see a **color** because of the way light hits something. Red, pink, orange, yellow, green, blue, purple, brown, gray, black, and white are all **colors.** A rainbow has many **colors** in it.

color

## come

**1.** Bill moves toward the kitchen. He **comes** toward it. He **came** to the kitchen to eat dinner.
**2.** The circus is in town once a year. It **comes** to town in the fall.
**3.** The juice at the store is kept in bottles and cans. But the apple juice only **comes** in bottles.

## complete

Every piece of the puzzle was put together. The puzzle was **complete**. There were no holes in it.

## computer

A **computer** is a machine that does work for people. A **computer** can help to find the answers to problems very quickly. Some **computers** can make pictures. There are special languages for **computers**.

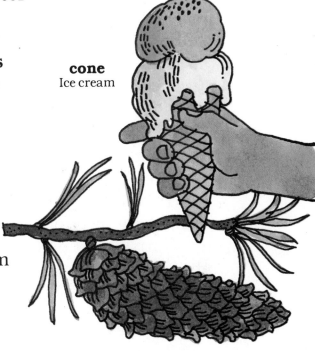

**cone**
Ice cream

## cone

**1.** A **cone** is an object with a circle at one end and a point at the other. Many people eat ice cream in **cones**.
**2.** A pine **cone** grows on a pine tree. Inside it are the seeds for new pine trees.

**cone**
Of a pine tree

## confuse

It is hard for Joan to remember the right names for some animals. Hippopotamus and rhinoceros are animals with strange names. She **confuses** them with each other. She **confused** them the first time she heard of them.

## consonant

A **consonant** is a kind of letter.  B, C, D, F, G, H, J, K, L, M, N, P, Q, R, S, T, V, W, X, and Z are always **consonants.**  Y is sometimes a **consonant.**

## contest

A game or a race is a **contest.**  Somebody must win or lose **contests.**

## continue

Baseball games stop in the rain.  They start again after the rain stops.  The games **continue** after the rain stops. One game stopped for two hours before it **continued.**

## control

David can make his kite rise and fall in the sky. He makes the kite do what he wants.  David **controls** the kite with a long piece of string. He **controlled** the kite until the wind stopped.

## cook

1. Andy heats a hamburger in a pan. He **cooks** it.  He **cooked** it for 15 minutes.
2. A **cook** is a person whose job is to make meals.  Most **cooks** do their work in restaurants.

cook

## cookie

A **cookie** is a small, flat cake.  It is sweet.  Most **cookies** are round.

## copper

**Copper** is a kind of metal.  **Copper** is used to make pennies.

## copy

**1.** Laura drew a picture of a tree.  Her picture looked just like a tree.  It was a **copy** of the tree.  Laura drew two **copies** of the tree to bring home.
**2.** Jane prints a poem on a piece of paper.  The poem is from a book.  Jane **copies** the poem from the book.  She **copied** the poem because she liked it.

## corn

**Corn** is a kind of vegetable.  It comes from the seeds of a tall, green plant.  **Corn** may be yellow or white.

**corn**

## corner

A **corner** is the place where two sides or two edges meet.  A square has four **corners.**

## correct

**1.** Jack got the right answer.  His answer was **correct.**
**2.** The teacher will read all the tests we took.  She will **correct** the tests by Wednesday.  We **corrected** our own answers on the last test.

## cost

Susan pays a dollar for a pumpkin. The pumpkin **costs** a dollar. The apples she bought yesterday **cost** more than that.

## costume

A **costume** is clothes people wear when they pretend to be someone or something else. Sometimes people wear **costumes** in plays. And **costumes** are an important part of Halloween.

costume

## cotton

**Cotton** is a kind of cloth. It is made from a plant. **Cotton** clothes are light. People wear **cottons** in the summer when it is very warm.

## could

Tom **can** whistle all of his favorite songs. He **could** whistle a short song with one breath. **Could** is a form of **can.**

## couldn't

**Couldn't** is a short way to write **could not.** Barry **couldn't** use the sled he got for his birthday. There was no snow on the ground.

## count

Kate adds together the pencils on her desk. She **counts** five pencils on her desk. She also **counted** four books.

## country

**1.** A **country** is a large area of land. The people in one **country** all live with the same laws. There are many **countries** in the world.
**2.** The **country** is an area of land far away from any city. The **country** is covered by forests, fields, and mountains.

## cousin

Your **cousin** is the child of your aunt or your uncle. Some people have lots of **cousins**.

## cover

**1.** Wendy puts a blanket over herself at night. She **covers** herself to keep warm. She **covered** herself with two blankets on winter nights.
**2.** A **cover** goes on the top or outside of things. Books have **covers**. Pots and pans have **covers**.

## cow

A **cow** is a kind of animal. It is large and likes to chew grass. Milk comes from **cows**.

**cowboy**

## cowboy

A **cowboy** is a person whose job is to take care of cattle. **Cowboys** usually ride horses.

## cowgirl

A **cowgirl** is a woman whose job is to take care of cattle. **Cowgirls** usually ride horses.

## crack

After the earthquake, the sidewalk had a crooked line through it. This **crack** was made when the ground under the sidewalk moved. Earthquakes make many **cracks** in streets and buildings.

## crash

**1.** The glass bowl fell off the table. It broke on the floor. The **crash** of the bowl was loud as it hit the floor. **Crashes** make a lot of noise.
**2.** The ball goes through the window. It breaks the window. It **crashes** through the window. The ball hit the floor of the gym after it **crashed** through the window.

## crawl

Babies move around on their hands and knees. They **crawl** from one place to another. My little brother **crawled** until he learned to walk.

## crayon

A **crayon** is a piece of colored wax. Liz draws pictures with her **crayons.**

## cream

**Cream** is a kind of food. It is part of the milk that comes out of a cow. Butter is made from **cream.**

## crocodile

A **crocodile** is a large animal. It looks like an alligator with a narrow jaw. **Crocodiles** live in swamps and rivers.

**crocodile**

## crooked

**Crooked** is the opposite of straight. Lightning looks like a **crooked** line in the sky.

## cross

**1.** A **cross** is a shape like this ( **+** ). Red **crosses** are seen in hospitals.
**2.** The horse runs from one side of the field to the other. It **crosses** the field. The horse **crossed** the field quickly.
**3.** The bridge was built over the river. It **crosses** the river near a town.

## crow

A **crow** is a kind of bird. It has black feathers. **Crows** like to eat corn.

## crowd

A **crowd** is a large group of people. There were **crowds** of people at the football game.

## cry

Babies make a lot of unhappy noises. They **cry** when they are hungry or thirsty. One baby **cried** for hours during the night.

## cup

A **cup** looks like a short glass with a handle. Many **cups** are made of clay.

**cup**

## curious

Kevin wonders what makes the snow fall and the wind blow. He is **curious** about the weather.

## curly

Megan's hair twists around in small circles. Her hair is **curly**.

## curve

**1.** The bottom of a U is a **curve.** It bends around and does not make a point. **Curves** have no corners.

**2.** The edge of a circle **curves** around until it meets itself. Bill drew an oval that **curved** around until it met itself too.

## cut

Ned divides a piece of paper with a pair of scissors. He **cuts** the paper into pieces. Yesterday he **cut** a picture out of the newspaper.

**cut**

# ABCDEFGHIJKLM
# NOPQRSTUVWXYZ

# Dd

## daisy

A **daisy** is a flower.
Most **daisies** are white.
They have yellow centers.

## dam

A **dam** is a wall built
across a river. The wall
keeps the water back.
A lake forms behind the **dam**.
People and beavers build
**dams**.

**daisy**

**dam**
To form a lake

**dam**
Made by a beaver

## damp

Eric washed his face. Then he dried it
with a towel. The towel became **damp**. It was
not very wet.

## dance

Mary and George listen to music. They move
their feet and hands with the music. They
**dance** until the music stops. Mary and
George **danced** with each other and
with their friends.

## dandelion

A **dandelion** is a flower.
**Dandelions** are round
and yellow.

## danger

Anything that can hurt people
is a **danger** to them.
Earthquakes, tornadoes,
and hurricanes are great
**dangers.**

## dark

**Dark** is the opposite of light.
It is **dark** outside at night.
Shadows are **dark.**

## date

A **date** is any one day.
July 7 is a **date.** So is
November 25. All days
have **dates.**

## daughter

A girl is the **daughter**
of her parents.
Some families have several **daughters.**

## day

1. A **day** is the time from one morning
to the next morning. There are 24 hours
in one **day.** There are seven **days** in a week.
2. **Day** is the opposite of night. The sun
shines during the **day.** **Days** begin
when nights end.

## dead

**Dead** is the opposite of alive. Most leaves die
in the fall. They are **dead** after the wind blows
them off the trees.

**dandelion**

D

## dear

Maggie wrote a letter to Susan. "**Dear** Susan," her letter began. People write **dear** to begin letters just as they say "Hello!" on the telephone.

## December

**December** is the last month of the year. It has 31 days.
**December** comes after November and before January.

## decide

Nora must choose a dress to wear to a party. She must **decide** which dress to wear. She **decided** to wear the blue dress.

## deep

**1.** It is very far from the boat to the bottom of the lake. The lake is **deep.**
**2.** Ben took a big breath after he ran in a race. He took several of these **deep** breaths.

## deer

A **deer** is an animal. It has soft fur. **Deer** have four legs. They can run very fast.

deer

## delicious

John thought his dinner tasted very good. He thought the food was **delicious.**

## dentist

A **dentist** is a kind of doctor. **Dentists** take care of people's teeth.

## describe

Cathy tells Dana about her new cat. She tells Dana what the cat looks like and feels like. Cathy **describes** the cat to Dana. She even **described** the little spots of white fur on its head.

## desert

A **desert** is a large area of land. Not much rain falls there. Sand covers most of the ground. Few plants or animals can live in **deserts.**

desert

## design

Diane draws pictures, circles, squares, and triangles on a piece of paper. She always draws them in a special order. She repeats this **design** all over the paper. Diane likes to draw **designs** on birthday cards.

## desk

A **desk** is a kind of furniture. It has a large flat top. People sit at **desks** to write.

## dessert

A **dessert** is a kind of food. It is the last part of a meal. Ice cream and fruit are **desserts.**

dessert

## diamond

**1.** A **diamond** is a jewel. It is very hard. **Diamonds** are found underground in rocks. They are as clear as glass.

**2.** A **diamond** is a shape. It has four sides. The top and bottom of a **diamond** are always corners. There are red **diamonds** on the cards we play games with.

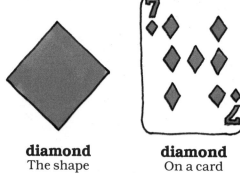

**diamond**
The shape

**diamond**
On a card

## dictionary

A **dictionary** is a book that explains what words mean. This book is a **dictionary.** The words in **dictionaries** are put in the same order as the alphabet.

## did

**1.** Our dog always carries the newspaper into the house. The dog **does** this every day. The dog even **did** it during the rain.

**2.** "Who wants some dessert?" Barbara asked. "I **did**," said Alex. "But I ate too much to want it now."

**Did** is a form of **do.**

## didn't

**Didn't** is a short way to write **did not.** Rick **didn't** feel well. He felt sick.

## die

People, animals, or plants that **die** are not alive. They will never think or move or grow again. The flowers in the garden **died** in the fall.

## different

**Different** is the opposite of same. A dog and an elephant are not the same shape or size. They do not sound alike either. Dogs and elephants are very **different** from each other.

## dig

Henry uses a shovel to make a large hole in the ground. He **digs** a hole in his garden. Henry **dug** the hole for some plants.

## dime

A **dime** is a coin. It will buy the same things as ten pennies or two nickels. Ten **dimes** will buy the same things as one dollar.

## dinner

**Dinner** is a big meal. People eat their **dinners** at noon or in the evening.

## dinosaur

A **dinosaur** was a huge animal. It lived millions of years ago. Some **dinosaurs** ate plants and some of them ate other **dinosaurs.** Some **dinosaurs** grew to be as big as three elephants put together.

dinosaur

## dip

John puts his toes in the lake. He **dips** his foot into the water to find out how warm it is. The water is very cold. After he **dipped** his foot into the water, he decided not to go in.

## direction

**1.** A **direction** is a way you can look or point or go. Oliver points at the moon. He points in the **direction** of the sky. North, south, east, and west are **directions**.

**2.** The **directions** for a game are a group of rules. They explain how the game should be played.

direction

## dirt

The ground is made of **dirt** and rocks. Rachel gets **dirt** on her clothes when she works in her garden.

## dirty

John has dirt all over himself. John is **dirty**.

## disappear

When a cloud covers the sun, the sun **disappears**. Nobody can see the sun then. The sun **disappeared** behind a lot of clouds while the rain fell.

## dish

A **dish** is something that people eat food from. It is flat and usually round. Michael washed all the **dishes** after his family ate supper.

## distance

It is four miles from the river to the town. The **distance** between the river and the town is four miles. Planes travel long **distances** in only a few hours.

## divide

**1.** Ken cuts an apple into two pieces. He **divides** the apple. Ken **divided** the apple in half.
**2.** When you **divide** a number, you find out how many of another number is in it. Frances **divided** 6 by 3. She got 2.

## do

**1.** Our dog always carries the newspaper into the house. The dog **does** this every day. The dog even **did** it during the rain. The dog will **do** it tomorrow too.
**2.** "Who wants some dessert?" Barbara asked. "I **do**," said John.

## dock

A **dock** is a place to tie up a boat. **Docks** stick out into a lake or the ocean from the shore.

dock

## doctor

A **doctor** is a person whose job is to help sick people get well. **Doctors** go to school for many years to learn all about medicine.

## does

**1.** Our dog always carries the newspaper into the house. The dog learned to **do** this a long time ago. The dog **does** it every day.
**2.** "Who wants some dessert?" Barbara asked. "John **does**," said Sue. "He's always hungry." **Does** is a form of **do**.

## doesn't

**Doesn't** is a short way to write **does not**. A turtle **doesn't** run as fast as a rabbit.

## dog

A **dog** is an animal. It has four legs, soft fur, and a tail. It makes a good pet. **Dogs** come in many different shapes, sizes, and colors.

dog

## doghouse

A **doghouse** is a small house for a dog. **Doghouses** are built outside in the yard.

## doll

A **doll** is a kind of toy. Most **dolls** look like babies or small children.

## dollar

A **dollar** is a kind of money. One **dollar** buys the same things as 100 pennies. **Dollars** are made of paper or of metal coins.

doll

## done

Our dog carries the newspaper into the house. The dog **does** this every day. The dog has **done** it this week, and he will **do** it next week. **Done** is a form of **do.**

## donkey

A **donkey** is an animal.  It has four legs and looks like a small horse. **Donkeys** have long ears.

## don't

**Don't** is a short way to write **do not.**  Birds **don't** fly north for the winter. They fly south.

## door

A **door** covers the place where you go into or out of a room. Barry opens the **door** to walk from one room to another. **Doors** are made of wood, metal, or glass.

**donkey**

## dot

A **dot** is a small, round spot. An "**i**" has a **dot** at the top of it. Some shirts and dresses are covered with **dots.**

## doughnut

A **doughnut** is a small, round cake. Many **doughnuts** have a hole in the center. Others are filled with jelly.

## down

**Down** is the opposite of up.
**1.** There is water at the bottom of the well. The water is **down** at the bottom of the well.
**2.** The goat went from the top of the mountain to the valley below. The goat went **down** the mountain.

## Dr.

**Dr.** is a word people use with a doctor's name.
**Dr.** Smith takes care of people when they are sick.

## dragon

A **dragon** is a monster in stories. It flies through the air and breathes fire. **Dragons** are almost as big as trucks.

dragon

## drank

Susan **drinks** juice from a glass. She **drank** it quickly.
**Drank** is a form of **drink.**

## draw

Laura makes pictures on paper. She **draws** the pictures with pencils, pens, or crayons. She **drew** a big picture of her brother. It looked just like him.

## drawn

Laura **draws** pictures with pencils and pens. She has **drawn** pictures with crayons too.
**Drawn** is a form of **draw.**

## dream

1. A **dream** is a story people imagine while they sleep. Nightmares are **dreams** that scare people.
2. Tina imagines herself in stories while she sleeps. She **dreams** about school, her friends, and her family. One night she **dreamed** she could fly.

## dress

1. A **dress** is a kind of clothes. The bottom of it is round. There are holes in the top of it for the head and the arms. Girls.and women wear **dresses.**
2. Abby puts clothes on her baby brother. She **dresses** him in the morning. Last night she **dressed** him in pajamas to go to sleep.

## drew

Laura **draws** pictures with pencils, pens, or crayons. She **drew** a big picture of her brother. It looked just like him.
**Drew** is a form of **draw.**

## drink

1. Susan puts juice into her mouth and swallows it. She **drinks** the juice from a glass. She **drank** it quickly.
2. Any liquid that people swallow is a **drink.** Water and milk are **drinks.**

**dress**

## drive

Tim's sister has a truck. She makes it go where she wants it to. She **drives** to work every day. She **drove** Tim to town yesterday.

## driven

Tim's sister **drives** to work every day. She has **driven** to work since she got her truck.
**Driven** is a form of **drive.**

## drop

**1.** A **drop** is a very small amount of a liquid. Many **drops** of rain fall during a storm.
**2.** Andy lets the cup fall. He **drops** the cup, and it breaks on the floor. When he **dropped** the cup, it was an accident.

## drove

Tim's sister **drives** a truck. She **drove** Tim to town yesterday.
**Drove** is a form of **drive.**

## drugstore

A **drugstore** is a store where people can buy medicine. Many **drugstores** also sell newspapers, food, drinks, ice cream, and other things.

## drum

A **drum** is an instrument. It is round on the top and the bottom. **Drums** are played to make beats for songs.

## drunk

Susan **drinks** a lot of juice. She has **drunk** three glasses of juice this morning.
**Drunk** is a form of **drink.**

drum

## dry

**1. Dry** is the opposite of wet. The sand in the desert has almost no water in it. The sand is **dry.**
**2.** Vicki covers her wet face with a towel. She **dries** her face with the towel. She **dried** her hands too.

## duck

A **duck** is a bird. It has a wide bill and short legs. **Ducks** like to swim on rivers and lakes.

## dug

Henry **digs** a hole in the garden with a shovel. He **dug** the hole for some plants.

**Dug** is a form of **dig**.

## dull

1. **Dull** is the opposite of sharp. The end of the pencil was smooth and flat. It was **dull**.
2. **Dull** is the opposite of exciting. The television show made Jan yawn. It did not interest her. She thought the show was **dull**.

**duck**

## during

The sky is blue while the sun shines. It is blue **during** the day.

## dust

**Dust** is tiny, soft pieces of dirt. The wind blows **dust** across the road.

ABCDEFGHIJKLM
NOPQRSTUVWXYZ

Ee

## each

**Each** means every one. Nancy picked berries off a bush. She picked them one at a time. She picked **each** berry carefully.

## each other

**Each other** means one and the other one. Joyce likes Karen. Karen likes Joyce. They are friends. They like **each other.**

## eagle

An **eagle** is a large bird. It has long wings. **Eagles** fly very high.

eagle

## ear

The **ear** is a part of the head. It is the part we hear with. People have two **ears.** One is on each side of the head.

## early

1. Bob gets up in the first part of the morning. He gets up **early** in the morning.
2. Mark was very tired. He went to bed before he usually does. He went to bed **early.**

ear

## earth

1. The **earth** is our world. We live on the **earth**.
2. Plants grow in the **earth**. Some animals dig holes in the **earth**. They live in the holes.

## earthquake

The ground shakes during an **earthquake**. Strong **earthquakes** make buildings fall down.

## east

**East** is a direction. The sun rises in the **east**. **East** is the opposite of west.

east

## easy

**Easy** is the opposite of hard. Jill knew all the answers on the test. The test was **easy** for her.

## eat

Chris puts food in his mouth. He chews and swallows it. Chris **eats** his food. He **ate** a big supper last night.

## echo

Jill shouted her name at the hills. She heard her name three times. The first **echo** of her name was loud. The other two **echoes** were soft.

## edge

1. An **edge** is the line or place where something ends. This page has three **edges**.
2. The **edge** of a knife is the sharp part that cuts.

## egg

1. An **egg** is a food. It has a thin, oval shell. George eats **eggs** for breakfast.
2. An **egg** is the place where a baby animal grows. Baby chickens hatch from **eggs**.

## eight

**Eight** is a number. **Eight** is written **8**.
7 + 1 is **8**.

## either

1. **Either** means one or the other. Frank had enough money to buy a basketball or a football. He could buy **either** of them.
2. **Either** means also. Jeff did not want to eat his dinner. He did not want to eat his dessert **either**.

## elbow

The **elbow** is a part of the arm. The arm bends at the elbow. **Elbows** are hard.

**elbow**

## electricity

**Electricity** is a kind of energy. It makes things work. **Electricity** makes lamps light up. It makes televisions, refrigerators, and many other things work.

## elephant

An **elephant** is a huge, gray animal. It has big ears and a trunk. No other animals that live on land are as big as **elephants**.

**elephant**

## else

Dennis wanted only a bicycle for his birthday. He did not want anything **else**.

## emerald

An **emerald** is a jewel. It is green. **Emeralds** are found in rocks.

## empty

The bottle has nothing in it.  The bottle is **empty.**

## end

**1.** The last letter of our alphabet is Z.  Our alphabet **ends** with Z.  The alphabet never **ended** with X.

**2.** Polly puts a straw into a glass.  One **end** of the straw is in the juice.  The other is in her mouth. Both **ends** of a straw look the same.

## energy

**Energy** is something that you cannot see but makes things work.   Heat and electricity are kinds of **energy.**

## engine

An **engine** is a machine.  It burns oil, gas, or wood to do work.  Cars have **engines** to make them go.

## engineer

An **engineer** is a person whose job is to drive a train. **Engineers** sit in the front of trains.

**engineer**

## enjoy

Paul likes to be at the beach.  He **enjoys** the beach every summer.  He especially **enjoyed** the beach on hot days.

## enough

We have five apples for five people.  We have **enough** apples for everyone.

## enter

Joe goes into his house from the garden.  He **enters** the house through the front door.  His dog **entered** the house through the kitchen door in the yard.

## especially

**Especially** means more than anything else.  Joanne likes to play soccer.  She likes to play tennis even more than soccer.  She **especially** likes to play tennis.

## even

**1. Even** is the opposite of odd.  Some **even** numbers are 2, 4, 6, 8, and 10.
**2.** The football field is flat and smooth.  The ground there is **even.**

## evening

**Evening** is a part of the day.  **Evening** begins when the sun sets in the west.  It ends when you go to bed at night.  The air is warm on summer **evenings.**

## ever

**Ever** means at any time.  Have you **ever** seen an elephant?

## every

The windows in the building are all dirty.  **Every** window needs to be cleaned.

## everybody

**Everybody** means each person.  All the people on the street opened their umbrellas when it started to rain.  **Everybody** stayed dry under their umbrellas.

## everyone

**Everyone** means everybody.  **Everyone** must eat and sleep to stay alive.

## everything

**Everything** means each thing.  When the snow fell, it covered **everything** on the ground.  The ground looked like a white blanket.

## everywhere

**Everywhere** means in every place.  Susan looked **everywhere** for her cat.  The cat could not be found.  But it returned home later.

## evil

**Evil** is the opposite of good.  The **evil** wizard in a story put everybody in a castle to sleep for a hundred years.

## excellent

Helen built a doghouse for her dog.  It was more than a very good doghouse.  It was an **excellent** one.

## except

Jason was the only student not in class today.  Everyone **except** Jason was in class.  He was sick at home.

## exciting

**Exciting** is the opposite of dull.  The fireworks made bright flashes of color in the sky.  They also made a lot of noise.  The fireworks were **exciting** to watch.

## excuse

Bill was not at school yesterday. Today he brought a note from his father. The note was an **excuse.** The **excuse** explained that Bill was sick yesterday. The teacher gave all the **excuses** to the school office.

## exercise

**1.** Nancy and Dennis run and play after school every day. They get a lot of **exercise.**
**2.** Matt is learning how to multiply. He did all the problems in his book. He did each **exercise.** The **exercises** were good practice.

## exit

An **exit** is the place where people go out of a room or building. Some **exits** are marked with signs.

## expand

When Alan blows air into a balloon, it **expands.** His balloon **expanded** until it broke.

**expand**

## expect

Jean thinks she will get many presents for her birthday. She **expects** to get many presents. She **expected** many presents last year, but she got only a few.

## expensive

Gold and silver cost a lot of money. They are very **expensive.**

## explain

Carole tells about things so that people can understand them. She **explains** things very well. She **explained** the rules of a game so we all could play.

## explore

Tom goes inside the cave. He wants to see what is in there. He **explores** the cave. Tom **explored** the cave with his friend Becky.

**explore**

## extra

Janice had six apples. She made a pie with five of them. She ate the **extra** apple.

## eye

An **eye** is a part of the head. People and animals have two **eyes.** They see with their **eyes.**

ABCDEF**F**GHIJKLM
NOP**Q**RSTUVWXYZ

# Ff

## face

**1.** The **face** is the front of the head. The eyes, the nose, and the mouth are all parts of the **face**. People who smile have happy **faces**.

**2.** The **face** of a clock shows what time it is. Clock **faces** have numbers on them.

**3.** Eric can see the front of the room. He **faces** the front. He **faced** the back before he turned around.

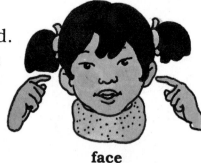

**face**
Of a person

## fact

A **fact** is something that is true. Elephants and whales are big. Zebras have stripes. These are **facts**.

**face**
Of a clock

## fair[1]

Peter and Dennis played on the swing. They took turns. They were very **fair** about the way they used the swing.

## fair[2]

A **fair** is a place where people go to have fun. Some people show things there that they have grown or made. At most **fairs** there are big machines to ride on and games to play.

## fairy

A **fairy** is a small magic person in stories. Some **fairies** are almost as big as people. Others are the same size as flowers.

**fairy**

# fall

**1. Fall** is a season. It comes after summer and before winter. Many **falls** begin with hot days and end with cold ones. Another name for **fall** is **autumn.**

**2.** An apple drops from the branch of a tree. It **falls** toward the ground. It **fell** until it hit the ground.

**3.** Paul goes to sleep at night. He **falls** asleep in his bed.

# fallen

**1.** An apple **falls** toward the ground. Many apples have **fallen** today.

**2.** Paul has **fallen** asleep early tonight. **Fallen** is a form of **fall.**

# family

A **family** is a father, a mother, and their children. **Families** usually live together in a house or an apartment.

# fancy

Janet gave a party for her friends. They all wore their best clothes. A band played music. It was a **fancy** party.

# far

**Far** is the opposite of near. It takes hours to travel from the top of the mountain to the valley below. The top of the mountain is **far** from the valley.

**fall**

## farm

A **farm** is a large area of land. People grow food or raise animals there. On some **farms** people do both.

farm

## farmer

A **farmer** is a person whose job is to work on a farm. **Farmers** start to work early in the morning.

## farther

Ben stands five feet away from George. Larry stands ten feet away from George. Larry is **farther** away from George than Ben is.

## fast

**Fast** is the opposite of slow. Rabbits move quickly. They are **fast** animals. Turtles are not **fast.**

## fat

A hippopotamus has a lot of extra weight. It is big and round. It is a **fat** animal.

## father

A **father** is a man who has at least one child. **Fathers** and mothers take care of their children.

fat

## favorite

Tom likes to read mystery stories more than anything else. Mystery stories are his **favorite** kind of book.

## fear

1. **Fear** is what people feel when they are afraid. Some people have **fears** of water, high places, or very small rooms.
2. Charlie does not know how to swim. He is afraid of the water. He **fears** it. He **feared** it because he could not swim. When he learns to swim, he will not **fear** the water.

## feather

A **feather** is part of a bird. Most birds are covered with them. **Feathers** are light and soft.

## February

**February** is a month of the year. It has 28 days for three years. Once every four years it has 29 days. **February** comes after January and before March.

## fed

Louise **feeds** her baby brother every morning. This morning she **fed** him cereal for breakfast. **Fed** is a form of **feed.**

## fee

A **fee** is an amount of money that has to be paid. We thought that the doctor's **fees** were too high.

## feed

Louise gives food to her baby brother. She **feeds** him every morning. She **fed** him cereal for breakfast.

**feather**

## feel

**1.** Jane is cold. She **feels** cold a lot during the winter. She **felt** cold outside in the snow.
**2.** Nick touches the point of his pencil. The point is sharp. It **feels** like the end of a needle.
**3.** Sandy is sick in the hospital. Her friends hope she will get well soon. They **feel** sorry for her.

## feet

**Feet** means more than one **foot.**
**1.** A person has two **feet.** Dogs have four **feet.**
**2.** Three **feet** is the same length as one yard.

## fell

**1.** An apple **falls** toward the ground from a tree. It **fell** until it hit the ground.
**2.** Paul **fell** asleep early last night.
**Fell** is a form of **fall.**

**feet**

## felt

**1.** Jane **feels** cold a lot during the winter. She **felt** cold outside in the snow.
**2.** The point of a pencil **felt** sharp to Nick.
**3.** Sandy's friends **felt** sorry for her when she was sick.
**Felt** is a form of **feel.**

## female

There are **female** and male people and animals in the world. Girls and women are **female** people.

## fence

A **fence** is like a wall. It is built to mark an area of land. **Fences** are made of wood, metal, or stone.

**fence**

## few

**Few** is the opposite of many. The trees in Amy's yard have only a **few** leaves on them in November.

## field

1. A **field** is a large, flat area of land. No trees grow on it. Farmers grow vegetables on **fields.**
2. A **field** is an area of land where a sport is played. Baseball and football are played on different kinds of **fields.**

field

## fight

1. Sometimes people **fight** when they can't agree. They may hit each other. They may just shout at each other. The two children **fought** over which television show to watch.
2. Sometimes a **fight** happens when people can't agree. **Fights** are usually not the best way to decide things.

## fill

Eric pours milk into a glass until it is full. He **fills** the glass with milk. He **filled** four glasses for his family.

## find

Matthew looks for his sneakers in his room. They are under his bed. He **finds** them under his bed. He also **found** three socks there.

### fine

**1.** Nora feels very good. She feels **fine.**
**2.** Patrick baked a cake. It looked and tasted very good. It was a **fine** cake.

### finger

A **finger** is a part of the hand. People have five **fingers** on each hand.

### finish

Joan works in her garden. After she is done, her hands are dirty. When she **finishes** in the garden, Joan washes her hands. Yesterday she **finished** early because it started to rain.

### fire

**Fire** is flame, heat, and light. It is what happens when things burn. Water puts out **fires.**

### fire engine

A **fire engine** is a kind of truck. Firefighters use it to help put out fires. **Fire engines** carry hoses, ladders, and firefighters to a fire.

**finger**

**fire, fire engine**

## firefighter

A **firefighter** is a person who puts out fires. Some **firefighters** put out fires in buildings.

## fireplace

A **fireplace** is a safe place to have a fire inside a house. It is a hole in a wall. There is a chimney for the smoke. **Fireplaces** are made of stone or metal. People burn wood in them.

**fireplace**

## fireworks

**Fireworks** make light, smoke, and noise in the sky. People shoot **fireworks** into the sky on small rockets.

## first

In the alphabet the letter A comes before all the others. It is the **first** letter of the alphabet.

## fish

1. A **fish** is an animal. It lives in the water. **Fish** have no arms or legs.
2. Tom tries to catch a **fish.** He **fishes** with a pole, a string, a hook, and a worm. Yesterday he **fished** all morning.

**fish**

**fish**
To catch fish

MATTHEW

## fisherman

A **fisherman** is a person who fishes. Good **fishermen** catch a lot of fish. Some **fishermen** go out to sea to catch fish to earn money to live on.

## fist

A **fist** is a hand that is closed up tight. Tim and Jenny knocked on the door with their **fists.**

## fit

**1.** Larry's new shirt is just the right size for him. It **fits** him well. The shirt **fitted** him until he grew too big for it.
**2.** Martha puts her clothes into a suitcase. She **fits** them all into it.

## five

**Five** is a number. **Five** is written **5.** 4 + 1 is **5.**

## fix

The refrigerator does not work. The food in it will not stay cold. After somebody **fixes** the refrigerator, it will work again. Paul **fixed** the refrigerator with some tools.

## flag

A **flag** is a large piece of cloth. It is covered with stripes, stars, or other symbols. Most countries have their own **flags.**

**flag**

## flame

A **flame** is part of a fire. It is shaped like a leaf. A **flame** is very hot. **Flames** change color and shape in fires.

## flash

1. A lighthouse shows a light for a moment. It **flashes** a light into the fog. It **flashed** a light many times during the night.
2. People see lightning for a moment. A **flash** of lightning looks like a crack in the sky. These **flashes** are very bright.

## flashlight

A **flashlight** is a kind of lamp. People carry **flashlights** with them at night.

## flat

Windows have no bumps in them. They have a smooth surface. Windows are **flat.**

## flavor

The taste of food is its **flavor.** Ice cream comes in many **flavors.**

## flew

1. Birds **fly** through the air. They **flew** from one tree to another.
2. The pilot **flew** the plane over the ocean.
3. Nancy **flew** to the mountains in a plane. **Flew** is a form of **fly.**

## float

Boats sit on top of the water. They **float** there. The boats that were tied to a dock **floated** in one place.

**flash**
To show light

**flash**
A flash of lightning

## flock

A **flock** of sheep is a group of sheep.
**Flocks** of animals live, travel, or
eat together.

## flood

In the spring
some rivers get
too big. They
pour over the dry
land near them.
A **flood** is the
water that goes
beyond these rivers over the
dry land. Dams are built to help
stop **floods.**

**flock**

## floor

A **floor** is the bottom of a room. People
walk on **floors.**

## flour

**Flour** is a powder. It is
made from wheat
or potatoes. **Flour** is
used to make bread
and cake. **Flours** are
usually white
or brown.

**floor**

## flower

A **flower** is a pretty
kind of plant. **Flowers**
come in many colors,
shapes, and sizes.
Roses and tulips are
**flowers.**

**flower**

## flown

**1.** Birds **fly** through the air. Some of them have **flown** hundreds of miles in one day.

**2.** The pilot has **flown** a plane over the ocean many times.

**3.** Nancy has **flown** in a plane on most of her vacations.

**Flown** is a form of **fly.**

## fly[1]

A **fly** is an insect. **Flies** have thin, clear wings.

## fly[2]

**1.** Birds move their wings to go through the air. They **fly** through the air. The birds **flew** from one tree to another.

**2.** A pilot makes a plane go in the right direction. A pilot **flies** a plane over land and sea.

**3.** Nancy travels in a plane on her vacation. She **flies** to different places.

## fog

**Fog** is a cloud that is near the ground. **Fogs** often gather near the ocean.

**fog**

## fold

Ned bends a piece of paper in half. He **folds** it.
He **folded** the paper to make a hat.

## follow

**1.** Ann walks behind Barbara. Ann **follows** Barbara through the forest. Ann **followed** Barbara to the lake.
**2.** Night comes after day. Night **follows** day.

## food

**Food** is something that people or animals eat. It helps keep them alive. Bread, hamburger, lettuce, and apples are all different **foods.**

## foot

**1.** The **foot** is a part of the body. It is at the end of the leg. The toes and the heel are part of the **foot.** People and birds have two **feet.** Cats and dogs have four **feet.**
**2.** A **foot** is an amount of length. One **foot** is 12 inches.

foot

## football

**1. Football** is a sport. It is played by two teams on a field. One team tries to throw, carry, or kick a football down the field. The other team tries to stop the first team.
**2.** The ball used in a football game is a **football.** It has an oval shape. **Footballs** are brown.

football

## for

**1.** The seal played with the ball between noon and one o'clock. The seal played with the ball **for** an hour.

**2.** Alice wanted the apple on the table. She reached **for** the apple with her hand.

**3.** The carpenter has a box to keep his tools in. It is a box **for** tools.

## forest

A **forest** is a large area with many trees and other plants. **Forests** can be found in warm or cold lands. Many different kinds of wild animals live in **forests.**

**forest**

## forget

Sometimes Ron does not remember to take his lunch to school. When he **forgets** his lunch, his friends share their lunches with him. Yesterday he **forgot** both his lunch and his books.

## forgot

Sometimes Ron **forgets** to take his lunch to school. When he **forgot** his lunch yesterday, his friends shared their lunches with him.

**Forgot** is a form of **forget.**

## forgotten

Sometimes Ron **forgets** to take his lunch to school. He has **forgotten** his lunch twice this week.

**Forgotten** is a form of **forget.**

## fork

A **fork** is a kind of tool to eat with. **Forks** are made of metal and have sharp points.

fork

## form

1. The **form** of an object is its shape. Clouds have many different **forms.**
2. A **form** is a kind of something. Ice is one **form** of water.
3. A word can have more than one **form.** "Did" is a **form** of "do."
4. Jenny shapes a piece of clay into an animal. She **forms** it into a dog. She **formed** another piece of clay into a giraffe.

## forward

Stuart's teacher called his name. He went to the front of the room. He went **forward** when his name was called.

## fought

The two children **fight** with each other. They **fought** over the basketball. **Fought** is a form of **fight.**

## found

Matthew **finds** his sneakers under his bed. He also **found** three socks and a book there. **Found** is a form of **find.**

## four

**Four** is a number. **Four** is written **4.** 3 + 1 is **4.**

## fox

A **fox** is an animal. It is about the size of a dog. It has four legs and a thick tail like a squirrel. Many **foxes** have red fur.

fox

## free

**1.** Bonnie has no plans for Saturday. She can do anything she wants. She is **free** for the day.
**2.** The toy store gave away some of its old toys. Nobody had to pay for them. The toys were **free.**

## freeze

Water changes into ice in cold weather. It becomes cold and hard. The water **freezes** and becomes ice. The lake **froze** in January.

## fresh

**1.** The bread was baked today. It is soft and delicious. The bread is **fresh.**
**2.** **Fresh** water has no salt in it. The water in rivers, ponds, and lakes is usually **fresh.**
**3.** The air on the mountain was clean and cold. It was **fresh** air.

## Friday

**Friday** is a day of the week. **Fridays** come after Thursdays and before Saturdays.

## friend

Molly and Anne like each other. They play together and share secrets. Anne is Molly's **friend.** Molly is Anne's **friend.** They have been **friends** for a long time.

## friendly

Elizabeth is a kind person who likes to meet other people. She is a **friendly** person.

## frog

A **frog** is a small animal. It has smooth skin and long, strong back legs. **Frogs** live near water. They like to eat flies.

**frog**

## from

**1.** Bob lived near the lake. Then he moved to a house near the mountain. Bob moved **from** a house near the lake to one near the mountain.

**2.** The swing and the ground were two feet apart. The swing hung two feet **from** the ground.

**3.** Alex knows the different kinds of apples by their tastes. He knows one kind of apple **from** another.

## front

**Front** is the opposite of back. The **front** of an old plane has a propeller on it.

## frost

**Frost** is very thin ice. It freezes on windows or the ground on cold fall mornings.

**frost**

## frown

**1.** Vicki is angry. Her nose is wrinkled. She **frowns** when she is angry. She **frowned** at her sister.

**2.** Sam is puzzled. He does not have a smile on his face. He has a **frown** on his face. **Frowns** make people look puzzled or angry.

## froze

Water **freezes** into ice in cold weather. The lake **froze** in January.
**Froze** is a form of **freeze.**

## frozen

Water **freezes** into ice in cold weather. The water in the lake has **frozen** into ice.
**Frozen** is a form of **freeze.**

## fruit

A **fruit** is a part of a plant. It has the seeds of the plant in it. Apples, oranges, and cherries are all **fruits.**

## fry

The chicken cooks in a pan filled with hot oil. It **fries** in the oil. It **fried** in half an hour.

## full

Ken filled a glass with milk. He filled the glass to the top. The glass was **full.** It would not hold any more milk.

## fun

Andy and his friends played games and sang songs at his birthday party. Everybody enjoyed the party. Everybody had **fun** there.

**fruit**

## funny

1. Jokes make people laugh. Jokes are **funny.**
2. Frank smelled something strange in the kitchen. It was a **funny** smell.

## fur

**Fur** is thick, soft hair. It covers many animals. Some animal **furs** are made into coats.

## furniture

Chairs, tables, beds, and desks are all kinds of **furniture.** People work on, eat on, and sleep on **furniture.** It is made of wood, metal, or other things.

## future

The **future** is a part of time. It is the part that has not happened yet. Tomorrow is part of the **future.**

A B C D E F G H I J K L M
N O P Q R S T U V W X Y Z

## gallon

A **gallon** is an amount of a liquid. Four quarts are the same as one **gallon.** Many **gallons** of milk are sold in supermarkets every day.

## game

A **game** is a way to play or have fun. Every **game** has rules. Some **games** are played with cards. All sports are **games.**

## garage

A **garage** is a building or a part of a building. Cars and trucks are parked or fixed in **garages.**
Many houses have **garages** in them. Some city **garages** can hold hundreds of cars.

**garage**

## garden

A **garden** is an area of land. People grow flowers or vegetables in **gardens.**

## gas

1. A **gas** is something that is so light that it fills up the thing that it is in. **Gases** usually have no color so we cannot see them. The air we breathe is a **gas.** Steam from very hot water is a **gas.**

2. **Gas** is sometimes several **gases** that are mixed together in a special way. It is used to make a fire that cooks food or keeps houses warm.

3. **Gas** is a liquid. It is burned in cars, trucks, and airplanes to make them move.

**garden**

## gate

A **gate** is a door in a fence. Some **gates** are built into stone walls too.

## gather

**1.** The birds in the tree all sit on the same branch. They **gather** together to sing. The birds **gathered** early in the morning.
**2.** Peter puts all his crayons together in a box. He **gathers** his crayons from all over his room and makes the room look neat.

**gate**

## gave

Melanie **gives** Peter a present on his birthday. She **gave** him a book last year. **Gave** is a form of **give**.

## geese

**Geese** means more than one **goose**. **Geese** fly south for the winter.

## get

**1.** Uncle Charles gives Laura a watch for her birthday. Laura **gets** a watch from him. She **got** a book last year.
**2.** Helen is hungry. She takes some fruit out of the refrigerator. She **gets** the fruit to eat.
**3.** Doug plays the piano a lot. He wants to become good. He **gets** good because he plays so much.
**4.** Carl arrives home a few minutes after he leaves school. When he **gets** home, he eats a sandwich.
**5.** Kelly lies in her bed. Then she stands up on the floor. She **gets** up from her bed.

# giant

1. A **giant** is a very large person in a fairy tale. Most **giants** are not friendly. But there are some **giants** that are kind.
2. **Giant** means very large. A tomato the size of a pumpkin would be a **giant** tomato.

# gibbon

A **gibbon** is a kind of monkey with very long arms. **Gibbons** live in the jungle. They swing from tree to tree. They use their long arms to do this.

# giddy

Alice and Bruce went on a ride at the fair. When they got off, they could hardly stand up. They were **giddy** from the ride.

# gift

**Gift** is another word for **present**. Larry gave David a basketball for his birthday. The basketball was a **gift**. David got a lot of **gifts** on his birthday. His parents gave him a bicycle.

# gigantic

Something that is **gigantic** is as big as a giant. We saw **gigantic** rocks when we got to the top of the mountain.

# giggle

1. A **giggle** is a short laugh. When Annette heard **giggles** in her closet, she knew her brother was in there.
2. Bill **giggles** when he tries to tell a funny story. One time he **giggled** so much that he didn't finish the story.

giant

## ginger

**Ginger** is the root of a plant. It is made into a powder and used in foods. It is also made into candy.

## gingerbread

**Gingerbread** is a kind of cake. Part of its flavor comes from ginger.

## giraffe

A **giraffe** is a tall animal. It has four legs and a very long neck. **Giraffes** are covered with brown spots. They like to eat the leaves off trees.

giraffe

## girl

A **girl** is a female child. **Girls** grow up to be women.

## give

Melanie buys a present for Peter. She **gives** it to him on his birthday. She **gave** him a book last year.

## given

Melanie **gives** Peter a present on his birthday. She has **given** him a present on each of his birthdays.
**Given** is a form of **give**.

## glad

Michael was happy the weather was good on Saturday. He was **glad** because he was at the beach.

## glare

Arthur looks at us with a very angry face. He **glares** at us. Last week he **glared** at us all day because his bicycle had broken for the third time.

girl

## glass

**1.** A window is made from **glass.** You can see through **glass.** It is hard and is easy to break. **Glass** is made from sand.

**2.** A **glass** is made to hold things. Nick drinks milk from a **glass.** He drinks two **glasses** of milk with his lunch.

**3.** People who do not see well wear **glasses.** The **glasses** fit over their noses in front of their eyes. The **glasses** help them see better.

glass

## glove

**1.** A **glove** is made of cloth, wool, or leather. It fits over a hand. People wear **gloves** to keep their hands warm in cold weather.

**2.** Susan wears a baseball **glove** when she plays baseball with her friends. It helps her catch the baseball.

**glove**
For cold weather

**glove**
For baseball

## glue

**Glue** is a thick liquid. When it dries, it holds things together.

## go

**1. Go** means to move from one place to another. Pam **goes** to school in the morning. She **went** to school late yesterday because the weather was bad.
**2.** Mary Ann and Joe must leave the party. They must **go** soon.
**3.** The new road was built through the forest. It **goes** through the forest.
**4.** Jean held the dog and would not let it move. Then she let it **go**.
**5.** Harry falls asleep at nine o'clock. He **goes** to sleep then.

## goat

A **goat** is an animal. It has horns, four legs, and short wool. **Goats** have a little piece of hair under their chins. **Goats** are raised for their milk and wool.

**goat**

## gold

**Gold** is a kind of metal. It is soft and yellow. It comes from underground or from streams. **Gold** is used to make jewelry, coins, and other things.

## goldfish

A **goldfish** is a kind of fish. It is small and orange. Some people keep **goldfish** in aquariums.

**goldfish**

## gone

**1.** Pam **goes** to school in the morning. She had **gone** to school with an umbrella.

**2.** Mary Ann and Joe are not at the party. They have **gone** home.

**3.** The new road has **gone** through the forest since it was built.

**4.** Harry has already **gone** to sleep tonight.

**Gone** is a form of **go.**

## good

**Good** is the opposite of bad.

**1.** Nancy liked the book she had just read. It was a **good** book.

**2.** Mark is happy. He smiles at everybody. He is in a **good** mood.

**3.** Penguins like cold weather. Cold weather is **good** for them.

## good-by

**Good-by** is the last word one person says to another when they talk. "**Good-by,**" Jill said to her friend over the telephone. Then she put the telephone back on the table.

## gooey

Honey sticks to things. It sticks to your peanut butter sandwich. It sticks to the spoon you use to get it out of the jar. It sticks to your fingers. Honey is **gooey.**

## goose

A **goose** is a bird. It looks like a large duck. **Geese** have long necks.

**goose**

## got

**1.** Laura **gets** a watch for her birthday from her Uncle Charles. Last year she **got** a book from him.
**2.** Helen was hungry. She **got** an apple and a pear out of the refrigerator.
**3.** Doug played the piano well. He **got** good because he played so much.
**4.** When Carl **got** home from school, he ate a sandwich.
**5.** Kelly **got** up from her bed late yesterday morning.
**Got** is a form of **get.**

## gotten

**1.** Laura **gets** a watch for her birthday. She has **gotten** it from her Uncle Charles.
**2.** Helen was hungry. She has **gotten** some fruit out of the refrigerator.
**3.** Doug plays the piano well. He has **gotten** good at it.
**4.** Carl has **gotten** home from school early today.
**5.** Kelly had **gotten** up from her bed as soon as she woke up.
**Gotten** is a form of **get.**

## grade

Penny spent her first year at school in the first **grade.** Now she is in the second **grade.** There are six **grades** in Penny's school.

## grain

**1.** Flour and cereal are made from **grain.** The seeds of wheat, corn, and rice are all **grains.**
**2.** A piece of sand is very small. It is a **grain** of sand.

**grain**
Wheat, corn, and rice

## gram

A **gram** is an amount of weight.
A penny weighs almost three **grams**.

## grandfather

The father of Judy's father or
mother is Judy's **grandfather.**
Judy has two **grandfathers.**

## grandmother

The mother of Bob's father or
mother is Bob's **grandmother.**
Bob has two **grandmothers.**

## grape

A **grape** is a kind of fruit. It is small
and round. It is not as small as a berry.
**Grapes** are green or purple.

grape

## grapefruit

A **grapefruit** is a large, round
fruit. **Grapefruits** look
like oranges, but they are not
as sweet.

## grass

Grass is a green plant. Many
people grow **grass** in their yards.
It also grows in fields, meadows,
and parks. There are different kinds
of **grasses.**

grass

## grasshopper

A **grasshopper** is a large insect. It has long
legs. **Grasshoppers** can jump several feet
in one hop.

## gray

Gray is a color. Elephants and
rain clouds are **gray.**

grasshopper

## great

**1. Great** means very large. There is a **great** old tree in Alan's yard.
**2. Great** means very important. Alice has pictures of all the **great** baseball teams.
**3. Great** means very good. Eric had a **great** time on his vacation.

## green

**Green** is a color. Leaves and grass are **green** in the summer.

## grew

**1.** Pumpkins **grow** from seeds to be fruits the size of basketballs. The pumpkins **grew** all summer.
**2.** Sharon **grew** up to become a teacher.
**Grew** is a form of **grow.**

## grin

A **grin** is a big smile. Beth cut a **grin** on her pumpkin for Halloween. She likes to cut crooked **grins.**

## ground

**Ground** is the rocks and dirt on the earth. Flowers and trees grow in the **ground.**

## group

People or things that are together are a **group.** Three **groups** of students went to the museum. Each **group** went on a different bus.

## grow

**1.** Pumpkins get big as the summer passes. They **grow** from seeds to be fruits as big as basketballs. The pumpkins **grew** all summer.
**2.** When Sharon **grows** up, she will be as big as her parents. Sharon **grew** up to become a teacher.

## grown

**1.** Pumpkins **grow** from seeds to be fruits the size of basketballs. The pumpkins had **grown** all summer.

**2.** Sharon had **grown** up to become a teacher. **Grown** is a form of **grow.**

## grown-up

A **grown-up** is an adult. My older brother is a **grown-up.** Our parents are grown-ups.

## guard

**1.** Martha watches over the picnic basket. She **guards** the food from ants and flies. Martha **guarded** all of the food she did not eat.

**2.** A **guard** is a person who watches over something. There are always **guards** in front of the palace.

## guess

**Guess** means to try to think of the answer to something. Sally does not know where her cat is. She **guesses** that her cat is up in a tree. Sally **guessed** wrong. Her cat was in the garage.

guard

## gym

A **gym** is a place where people play games and do exercises. Many schools have large **gyms.**

A B C D E F G H I J K L M
N O P Q R S T U V W X Y Z

# Hh

## habit

Every night George has a glass of milk before he goes to bed. It is a **habit** he has. There are other things that George does often. Those **habits** have to do with school work and sports.

## had

Tom and Lucy **have** four quarters in the jar. They only **had** two quarters last week. **Had** is a form of **have**.

## hadn't

**Hadn't** is a short way to write **had not**. James wanted to see a movie, but he **hadn't** washed the dishes yet.

## hair

David has straight **hair** on his head. After he brushes his **hair** he finds a few **hairs** on the brush.

## haircut

Jeremy goes to a barber shop when his hair gets too long. The barber gives him a **haircut**. Now Jeremy's hair is short. Barbers give **haircuts**.

**haircut**

## half

Jim cuts a pie into two pieces. Each piece is the same size. Each piece is **half** of the pie. Jim's family will eat both **halves** of the pie this afternoon.

## hall

A **hall** is an area inside a building. People use **halls** to walk through the building. Rooms are built next to **halls**.

**half**

# Halloween

**Halloween** comes on the last day of October. Children dress up in funny clothes or costumes on **Halloween.** Then they collect candy in their neighborhood.

Halloween

## halves

**Halves** means more than one **half.** Jim cut the pie into two **halves.**

## ham

**Ham** is a kind of meat from pigs. Many people buy **hams** at the supermarket.

## hamburger

1. **Hamburger** is a kind of beef. Many people eat **hamburger** in a round, flat shape.
2. A **hamburger** is a sandwich. It is made with **hamburger** and two pieces of bread or a roll. Donald puts ketchup on his **hamburgers.**

## hammer

A **hammer** is a tool. It has a long handle and a heavy metal top. It is shaped like a T. **Hammers** are used to put nails into wood.

## hamster

A **hamster** is a small animal. It is covered with fur and has a short tail. Many people keep **hamsters** as pets.

## hand

A **hand** is a part of the body. The **hands** are at the end of the arms. People use their **hands** to hold things. The fingers, the thumb, and the palm are parts of the **hand.**

## handkerchief

A **handkerchief** is a piece of cloth. Charlie puts a **handkerchief** over his nose when he sneezes. Many **handkerchiefs** are white.

## handle

**1.** Susan held the suitcase by its **handle.** The **handle** was on the top of the suitcase. **Handles** make many objects easy to hold.
**2.** Ellen picks up a needle. She **handles** it with care because it is sharp. She **handled** the needle while she sewed a dress.

**hammer**

**hamster**

**handle**

125

# hang

A bat **hangs** onto a branch with its feet. But the bat is below the branch. It **hangs** upside-down. The bat **hung** there while it slept.

# happen

The magician puts his hand into his hat. What will come next? What will **happen?** The magician pulled a rabbit out of his hat. That is what **happened**.

# happy

**Happy** is the opposite of sad. Donna smiled and laughed at the party. Donna was **happy.**

# hard

**1. Hard** is the opposite of soft. Phillip hit his hand on a rock. The rock was **hard.** Now his hand hurts.

**2. Hard** is the opposite of easy. Beth walks home from school in the rain. She does not have an umbrella. It is **hard** for her to stay dry without an umbrella.

# hardly

The clouds covered the sky today. Most of the time we could not see the sun. We **hardly** saw it all day.

# has

Tom and Lucy **have** four quarters. Lucy **has** two quarters. Tom **has** two quarters.

**Has** is a form of **have**.

# hat

**hat**

A **hat** is something you wear on your head. **Hats** come in many different shapes and sizes.

## hatch

When a baby chicken breaks out of its egg, it
**hatches.** Five chickens **hatched** from their eggs
yesterday.

hatch

## haunt

A ghost lives in an old house in the forest. He
scares everybody who comes near the house. He
**haunts** the house. The ghost **haunted** the house
for a hundred years.

## have

The four quarters in the glass jar belong to Tom
and Lucy. Tom **has** two of them and Lucy **has**
two of them. They **have** four quarters
in the jar. They only **had** two quarters
last week.

## haven't

**Haven't** is a short way to write **have not.**
Maggie and Lynn **haven't** been to the circus yet.

## hawk

A **hawk** is a bird. It has a short, curved
beak. **Hawks** eat small animals.
They can fly high in the sky.

## hay

**Hay** is a kind of tall grass
that has been cut and dried.
Horses and cows eat **hay.**

hay

## he

Edward is a boy. **He** is a male person.

## head

**1.** The **head** is a part of the body. People's eyes and ears are part of their **heads.**
**2.** Eric will be the first person to get into the movie theater. Everybody else stands behind him. Eric is at the **head** of the line.

## heal

Jack cut his finger with a sharp knife. When it **heals** the cut won't hurt and it won't show. His finger **healed** in a week.

## healthy

Alice feels good and she is not sick. She is **healthy.**

## hear

Pam sees with her eyes. She **hears** with her ears. Pam **hears** the thunder during a storm. She **heard** it after she saw the lightning.

## heard

Pam **hears** the thunder during a storm. She **heard** it after she saw the lightning. **Heard** is a form of **hear.**

## heart

**1.** The **heart** is a part of the body. It makes the blood move through the body. The **heart** is in the chest. We can feel the beat of our **hearts** when we run.
**2.** A **heart** is a shape that means love.

**heart**

## heat

**1. Heat** makes you feel warm. The **heat** from the sun helps plants grow. The **heat** in the oven cooks the food inside it.
**2.** The fire in the fireplace **heats** the room. It **heated** the room all night.

## heavy

**Heavy** objects are hard to lift. Bowling balls are **heavy.** Feathers are not.

**heavy**

## heel

**1.** The **heel** is a part of the body. It is the round back part of the foot. When Karen puts her feet into shoes, her **heels** go in last.
**2.** The **heel** is part of a shoe. It fits under the back of the foot. Some shoes have high **heels.**

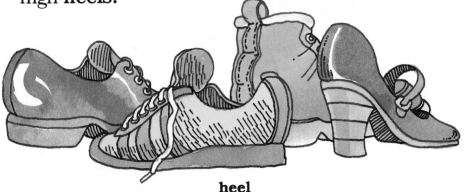

**heel**

## height

The mountain is 1,000 feet tall. Its **height** is 1,000 feet. Bill is not afraid of **heights.**

## held

**1.** Judy **holds** the baby against her shoulder. The baby **held** the blanket in his hands.
**2.** The car only **held** four people.
**Held** is a form of **hold.**

## helicopter

A **helicopter** is a machine. It carries people through the air. It does not have wings like a plane. **Helicopters** have propellers on top of them. They turn around and around. They lift the **helicopter** off the ground.

helicopter

## hello

**Hello** is a word people say to each other. When most people pick up a telephone that rings, they say "**Hello.**"

## help

The rock in the garden is too heavy for Janet to lift by herself. Her mother **helps** her lift it. They lifted the rock together. Her mother **helped** Janet lift three rocks.

hen

## hen

A **hen** is a bird. Female chickens grow up to be **hens**. **Hens** lay eggs.

## her

**1.** Tony saw Rose at a party. He told **her** that she looked pretty.
**2.** Emily likes the blue dress she wears to school. It is **her** favorite dress.

## here

**Here** means in this place. "Where are you, Julie?" her mother called out the window. "I'm **here** in the garden," said Julie.

## hers

Kate bought a kite to use in the park. She owns the kite. It is **hers.**

## herself

Linda put on a Halloween mask. Then she looked in the mirror. Linda saw **herself** in the mirror. She had a mask on her face.

## hid

Squirrels **hide** nuts in the fall. They **hid** the nuts where they could find them.
**Hid** is a form of **hide.**

## hidden

Squirrels **hide** nuts. They have always **hidden** them.
**Hidden** is a form of **hide.**

## hide

A squirrel dug a hole in the ground. It put a nut in the hole. Squirrels **hide** nuts in the fall. They **hid** the nuts where they could find them.

## high

**High** is the opposite of low.
**1.** The hawk flew through a cloud. It was far from the ground. The hawk flew **high** in the sky.
**2.** Gold costs a lot. It has a **high** price.

## hill

A **hill** is a big bump in the ground.
Mountains are large
**hills.**

hill

## him

Rose met Tony. She was glad to see **him.**

## himself

Michael put on a baseball cap. Then he looked
in the mirror. Michael saw **himself**
in the mirror. He had a hat on his head.

## hippopotamus

A **hippopotamus** is a large
animal. It is the size
of a small car.
They have short legs
and huge mouths.
They live near lakes
or rivers.

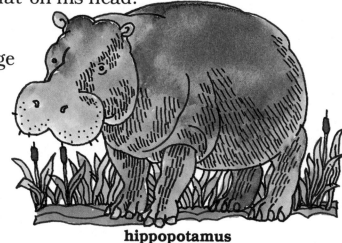

hippopotamus

## his

That bicycle belongs to Alex. It is **his** bicycle.

## hit

Bill swings his bat at the baseball. He **hits** the
baseball with the bat. Yesterday he **hit** the ball
across the field.

## hive

A **hive** is a home for bees. Bees make honey in their **hives.**

## hobby

Eliza collects stamps from different places. It is a **hobby** of hers. Some of her friends have other **hobbies.**

hive

## hockey

**Hockey** is the name of two sports. In **field hockey** two teams hit a ball across a field with sticks. In **ice hockey** two teams hit a puck across a patch of ice. In **ice hockey** the teams wear skates. Both **hockeys** are played at schools.

**hockey**
Field hockey

**hockey**
Ice hockey

## hold

**1.** Judy carries a baby in her arms. She **holds** the baby against her shoulder. The baby **held** a blanket in his hands.
**2.** The car has space for four people. The car is small. It will only **hold** four people.

## hole

**1.** A **hole** is an empty space in an object. Many doughnuts have **holes** in the middle of them.

**2.** Joe dug some dirt out of the ground. The space he has made in the ground is a **hole.**

## holiday

A **holiday** is a very special day. On some **holidays,** we think about important people from the past. On other **holidays,** we remember important things that have happened.

hole

4th of JULY PA

holiday

## hollow

A **hollow** space is an empty space inside an object. Basketballs and chimneys are **hollow.**

**hollow**

## home

**Home** is the place where people or animals live. Most people have **homes** in houses or apartments.

## honest

**Honest** people are fair to everyone. They do not lie or take things that belong to somebody else.

## honey

Bees make **honey.** It is a thick liquid. **Honey** is sweet like syrup. Bears like **honey** even more than people do.

## hook

A **hook** is a curved piece of metal. Some **hooks** are used to catch fish. Others are used to hang up clothes.

## hop

1. The rabbits jump and jump and jump across the field. They **hop** across it quickly. The rabbits **hopped** across the field to the stream.

**hook**

2. The rabbits each took one long **hop** over the stream. Their **hops** all ended in the mud on the other side.

## hope

Tim does not want it to rain on Saturday. The school picnic will not be much fun if it rains. He **hopes** Saturday will be sunny. He also **hoped** there would be good food to eat.

## horn

**1.** A **horn** is part of an animal's body. A unicorn in a fairy tale has only one **horn** on its head. Bulls and goats have two **horns** on their heads.
**2.** A **horn** is an instrument. Emily blows into her **horn** to make music.
**3.** A **horn** is something that makes a loud noise. The truck's **horn** told us to get out of the street.

**horn**
Of a bull

**horn**
An instrument

## horse

A **horse** is a large animal. It has a long head and four legs. **Horses** can run fast.

## horseshoe

A **horseshoe** is a piece of iron. It is shaped like a U. Horses wear **horseshoes.**

**horse**

**horseshoe**

## hose

A **hose** is a tube made of rubber or cloth. Firefighters use **hoses** to carry water.

## hospital

A **hospital** is a building. Doctors and nurses take care of sick people there. Many **hospitals** have hundreds of rooms.

## hot

**Hot** is the opposite of cold. It means very warm. Fires are **hot.**

## hot dog

A **hot dog** is a kind of food. It is made from meat and other things. People often eat **hot dogs** in long, thin rolls. **Hot dogs** taste good with mustard.

## hotel

A **hotel** is a big building. It has a lot of rooms. These rooms are rented to people who are away from home. Big cities have many **hotels.**

## hothouse

A **hothouse** is a special house to grow plants in. **Hothouses** usually have glass roofs. The air inside is always warm.

## hour

An **hour** is an amount of time. One **hour** has 60 minutes in it. There are 24 **hours** in one day.

hose

hot dog

## house

A **house** is a building. People
live there. Any number of people
can make their home
in a **house**. Some **houses** are
divided so that two or three
families can
live in them.

**house**

## how

**1.** Penny wanted to know the way to make a
rabbit come out of her hat. "**How** did you make
that rabbit come out of your hat?" she asked the
magician.
**2.** Jeff wanted to know the price of the jacket
in the store. "**How** much does that
jacket cost?" he asked.

## however

The ghost stopped in front of a door.
The door was locked. The ghost,
**however,** did not need a key. It just
went through the door like smoke
through a fence.

## huge

**Huge** means very big. Jim made
himself a sandwich with turkey, ham,
cheese, lettuce, tomato, and three pieces
of bread. This was a **huge** sandwich.

**huge**

## human

Anything **human** is about people. A **human** body is the body of a person.

## hump

A **hump** is a bump on an animal. Camels have either one or two **humps** that stick up out of their backs.

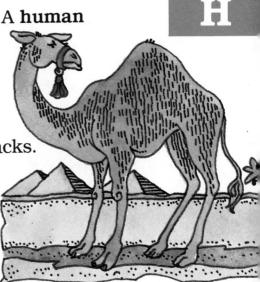

**hump**

## hundred

A **hundred** is a number. It is written **100.** It takes ten tens to make one **hundred.** There are **hundreds** of pages in this book.

## hung

The bat **hangs** upside-down from the branch. It **hung** there while it slept. **Hung** is a form of **hang.**

## hungry

Betsy wants to eat. She has had nothing to eat all day. Her stomach is empty. Betsy is **hungry.**

## hunt

Sam can never find his shoes in the morning. He looks for them all over his room. He **hunts** everywhere. Yesterday Sam **hunted** until he found his shoes under the bed.

## hunter

A **hunter** is a person or animal who hunts. Tigers have to be good **hunters** so they can eat. Penny's cat is not a good **hunter.** It chases squirrels and mice, but they always get away.

## hurry

Adam runs to school. He **hurries** because he is late. He **hurried** to school every morning this week.

## hurt

**1.** Jane's back got very red at the beach. Now it **hurts** when she touches it. Her skin **hurt** for two days the last time it got red.

**2.** Nora falls on the steps. She **hurts** her knee when she falls. She cries because her knee feels bad.

## husband

A **husband** is a married man. He is the **husband** of the woman he married. **Husbands** and wives are married to each other.

## hut

A **hut** is a very simple house. Most **huts** are built in warm places. Some of them have grass roofs.

**hut**

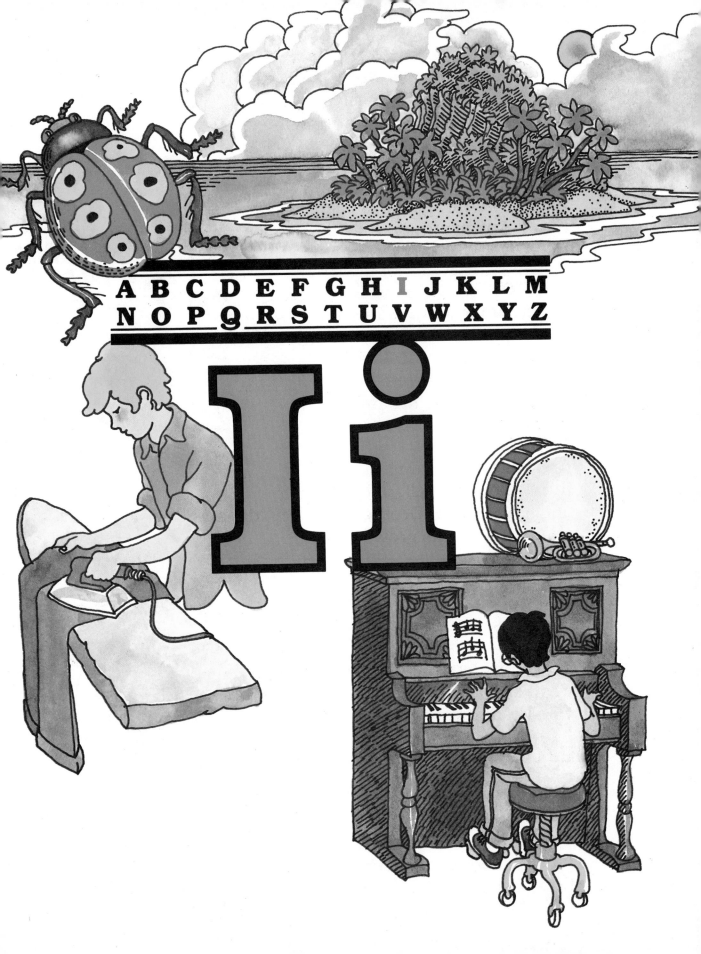

A B C D E F G H **I** J K L M
N O P Q R S T U V W X Y Z

I i

## I

I and me are the two words people use to write or to speak about themselves. The brown coat belongs to me. I wear it during the winter to keep me warm.

## ice

Ice is frozen water. It is cold and hard. Ann skates on the **ice** that covers the pond.

## ice cream

Ice cream is a frozen dessert. It is made from cream, eggs, sugar, and different flavors. Vanilla and chocolate are two flavors of **ice cream.**

ice

## idea

An **idea** is something that you think of by yourself. It can be about anything. George has many **ideas** about how plants grow.

## if

**1.** Jack may decide to go to the store. If he goes, he will buy a loaf of bread.
**2.** Wendy does not know what the weather will be like today. She wonders **if** it will snow.

## ill

Richard feels sick. He is not healthy. He is **ill.**

ice cream

## I'll

I'll is a short way to write **I will.** I'll eat breakfast after I get dressed.

## I'm

I'm is a short way to write **I am.** **I'm** almost as tall as my brother.

## imagine

Laura thinks about summer as the snow falls outside her window. She **imagines** herself on a beach. Laura **imagined** this a lot during the winter.

## important

1. Tracy wants very much to win a race at school. It is **important** to her to win a race.
2. The king and queen had a lot of power. They were very **important** people.

## impossible

Anything that is **impossible** cannot happen. It is **impossible** for fish to speak.

## in

1. Andy lives inside a small house. His room is **in** the house.
2. Leaves are green during the summer. They are green **in** July and August.
3. Sheila held the fork **in** one hand and the knife **in** the other.

## inch

An **inch** is an amount of length. One foot is 12 **inches** long.

inch

## ink

Ink is a liquid that people write with. When Abby writes with a pen, **ink** comes out of it. Blue and black **inks** are used in most pens.

## insect

An **insect** is a tiny animal. It has six legs. Some **insects** can fly. Flies, ants, butterflies, grasshoppers, and bees are all **insects**.

## inside

**Inside** is the opposite of outside.
1. Donna goes into her house through the front door. She goes **inside** to get warm.
2. The **inside** of a basketball has air in it.

insect

## instant

An **instant** is a very short amount of time. **Instants** are almost too short to notice. The lightning flashed for an **instant**.

## instead

Melissa wanted to eat an orange. But she only found pears in the refrigerator. So she ate a pear **instead**.

## instrument

An **instrument** makes music. Pianos, trumpets, drums, violins, and trombones are all **instruments**.

instrument

## interest

**1.** Eliza likes to read about animals.  She has an **interest** in them.  Eliza has many other **interests.**

**2.** Al is curious about kites.  They **interest** him. They **interested** him the first time he saw one.

## into

**1.** The worm crawled inside the apple.  It crawled **into** the apple.

**2.** Caterpillars become butterflies.  They change **into** butterflies.

## invent

Tom makes a machine that nobody has ever made before.  He **invents** the machine. Tom **invented** a lot of machines.

## invention

The thing that a person invents is an **invention.**  Cars and radios are important **inventions.**

## invite

Tina asks her friends to come to her house for supper.  She **invites** them to come.  Tina **invited** three other friends to the same supper.

## iron

**1. Iron** is a kind of metal.  It is hard and gray.  Many strong things are made of **iron.**

**2.** An **iron** is used to take the wrinkles out of clothes. **Irons** are flat on the bottom. They get very hot.  Some **irons** make steam.

iron

## is

John just had a birthday.  Now he **is** seven years old.  Most of his friends **are** seven too. John **was** six until his last birthday.

**Is, are,** and **was** are forms of **be.**

## island

An **island** is an area of land.  It sticks up out of a river, a pond, a lake, or an ocean.  **Islands** have water all around them.

## isn't

**Isn't** is a short way to write or say **is not.**  The sun **isn't** dark.

## it

1.  A tulip is pretty.  **It** is pretty.
2.  Snow falls in the winter.  **It** is possible that snow will fall many times this winter.

## itch

Tony's arm **itches** where the mosquito bit him. As long as it **itched,** Tony scratched it.

## its

The dog has a tail.  It played with **its** tail.  Then it played with **its** shadow.  The dog played alone.

## it's

**It's** is a short way to write **it is.**  **It's** fun to go to the circus.

## itself

The swan looked down at the lake.  It saw a swan in the water.  The swan saw **itself** in the water.

A B C D E F G H I J K L M
N O P Q R S T U V W X Y Z

Jj Kk

### jacket

A **jacket** is a short, light coat. Sarah wears a **jacket** to the summer fair. She has several **jackets** in her closet.

### jail

A **jail** is a kind of building. People who break the law must live there. **Jails** have bars across the windows.

### jam

**Jam** is a kind of food. It is made from fruit and sugar cooked together. **Jams** are thick and sweet.

### January

**January** is the first month of the year. It has 31 days. **January** comes after December and before February.

### jar

A **jar** looks like a fat bottle. Jam, jelly, and cookies are kept in glass **jars.**

jacket

jar

jail

## jaw

A **jaw** is a part of the body. It is a bone at the bottom of the face. When people speak, their **jaws** move.

## jelly

**Jelly** is a kind of food. It is made from fruit juice and sugar boiled together. Grape and apple **jellies** taste good with peanut butter.

## jet

A **jet** is a kind of plane. Its engines do not use propellers. Other planes cannot fly as fast as **jets.**

## jewel

A **jewel** is a kind of stone. Light can pass through it. Diamonds, emeralds, and rubies are **jewels.**

jaw

jewel

jet

## job

1. Barbara is a carpenter. This is her **job.**
Donald is a teacher. That is his **job.** People get
**jobs** after they finish school. They work
at their **jobs** to make money to live.
2. Ned must cut the lawn around his house.
That is his **job** for the day.

## jog

George runs every day. He does not race. He
**jogs** for exercise. Yesterday he **jogged** five miles
along the beach.

## join

1. Nancy puts the two pieces of paper together
with glue. She **joins** them together with glue.
Nancy **joined** several pieces of paper together.
2. Dan wants to belong to a basketball team. He
**joins** the school basketball team. Now he plays
with his team against other teams.

## joke

A **joke** is a short, funny story. **Jokes**
in the form of questions are riddles. Good **jokes**
make people laugh.

## jolly

Kevin's family has a lot of fun. They laugh a
lot. Kevin's friends like to visit this **jolly** family.

## joy

**Joy** is something people feel when they are very
happy.

## juggle

Karen can throw and catch three or more things
at the same time. She **juggles** very well.
Yesterday she **juggled** four tennis balls and
didn't drop any.

## juice

**Juice** is a kind of liquid. It is made
from the water in different foods. People drink
**juices** made from oranges, apples, grapes, and
tomatoes.

## July

**July** is a month of the year. It has 31 days.
**July** comes after June and before August.

## jump

**1.** Lily goes up into the air to catch the ball. Lily
**jumps** to catch it. She **jumped** very high.
**2.** The cat starts from the roof of the house,
goes into the air, and comes down on its feet.
The cat **jumps** from the roof of the house
to the ground.
**3.** The grasshopper goes up into the air and
moves three feet across the sidewalk. The
grasshopper **jumps** three feet.

jump

## June

**June** is a month of the year. It has 30 days.
**June** comes after May and before July.

# jungle

A **jungle** is an area of land. A lot of huge plants and trees grow there. Lions, tigers, monkeys, and snakes live in **jungles.**

jungle

# junk

**Junk** is something that nobody wants. Broken furniture and old cars are kinds of **junk.**

# just

**1.** School starts at nine o'clock. Cathy got there one minute early. She got to school **just** before her class began.

**2.** George wanted to buy a kite. It cost two dollars. He had two dollars. He had **just** enough money to buy the kite.

## kangaroo

A **kangaroo** is a large animal. **Kangaroos** move very fast when they jump on their strong back legs. **Kangaroo** mothers carry their babies in a pocket in front of their stomachs.

## keep

**1.** Joe got many presents for his birthday. He will not give them away. He will **keep** them. Joe **kept** every present except a shirt that was too big for him.

**2.** Kerry must be quiet while her baby sister sleeps in the afternoon. Kerry must **keep** quiet.

**3.** Mark puts his clothes in a closet. He **keeps** them there until he wears them.

**kangaroo**

## kept

**1.** Joe will **keep** the presents he got for his birthday. He **kept** every present except a shirt that was too big for him.

**2.** Kerry **kept** quiet while her baby sister slept in the afternoon.

**3.** Mark **kept** his clothes in his closet.

**Kept** is a form of **keep.**

## ketchup

**Ketchup** is a very thick, red liquid. It is made from tomatoes and other things.

## kettle

A **kettle** is a kind of pot. Water is boiled in **kettles.**

## key

1. A **key** is a piece of metal. It opens a lock. Many people have **keys** to open the doors of their homes and cars.
2. A **key** is part of a piano. It is either white or black. When Peggy plays the piano, she puts her fingers on the **keys.**

## keyhole

A **keyhole** is a hole in a lock. Keys fit into **keyholes.**

## kick

Bob hits the football with his foot. He **kicks** the ball as far as he can. He **kicked** the ball across the field.

key

## kill

The cold weather makes the flowers in Peter's garden die. The weather **kills** the flowers. After the weather **killed** the flowers, Peter dug up the garden for next spring.

## kilometer

A **kilometer** is an amount of length. 1,000 meters is one **kilometer.** Five **kilometers** is about the same distance as three miles.

## kind

1. John is a friendly person. He likes to help other people. He is a **kind** person.
2. Lettuce is a **kind** of vegetable. Carrots and potatoes are other **kinds** of vegetables.

## kindergarten

**Kindergarten** is a class in school. It is the year before first grade. In most **kindergartens** the students go to school for just half a day.

## king

**1.** A **king** is a man who rules a country. **Kings** often rule for as long as they live.

**2.** A **king** is the name of a card used in some games. The **king** card has a picture of a **king** on it.

## kingdom

A **kingdom** is a country where a king or a queen rules. In fairy tales princes and princesses from different **kingdoms** often love each other.

king

## kiss

Lisa touched her father's cheek with her lips. She gave him a **kiss** before she fell asleep. Lisa gives her father many **kisses.**

## kit

**1.** A **kit** is a set of things that have to be put together. Model cars and airplanes come in **kits.**

**2.** A **kit** is a small set of tools.

## kitchen

A **kitchen** is a kind of room. People cook and keep food there. Sometimes they eat there too. Refrigerators and ovens are used in **kitchens.**

**K**

### kite

A **kite** is a kind of toy. It is made of sticks of wood. The sticks are covered with paper or cloth. **Kites** are carried into the sky by the wind.

**kite**

### kitten

A **kitten** is a young cat. **Kittens** like to play in paper bags.

### knee

The **knee** is a part of the leg. It bends like the elbow. People bend their **knees** when they sit down.

### knew

1. Bonnie **knows** how to spell rhinoceros. She **knew** how to spell the names of many animals.
2. Tony **knew** Ben because they were in the same class. **Knew** is a form of **know**.

**knee**

## knife

A **knife** is a kind of tool. It is a sharp piece of metal. Oliver uses a **knife** to cut the food on his dish. **Knives** come in many different sizes.

**knife**

## knight

Hundreds of years ago a **knight** was a person who traveled through different kingdoms. **Knights** helped people in danger. **Knights** also fought each other in contests.

## knives

**Knives** means more than one **knife**. **Knives** come in many different sizes.

## knock

Ellen hits a door with her fist. She **knocks** on the door. She **knocked** three times before her friend Beth opened it.

**knight**

## knot

Nancy tied a ribbon in a bow.
In the middle of the bow was
a **knot**. The **knot** is the place
where the ribbon is pulled tight.
**Knots** can be made with string
and rope too.

## know

**1.** Bonnie has learned how to spell
rhinoceros. She **knows** how
to spell it. Bonnie **knew** how
to spell the names of many animals.
**2.** Tony meets Ben in his class at school. Tony
**knows** Ben after they meet.

## known

**1.** Bonnie **knows** how to spell rhinoceros. She
has **known** how to spell it since she saw the
word in a book.
**2.** Tony has **known** Ben for a year.
**Known** is a form of **know**.

knot

ABCDEFGHIJK**L**M
NOPQRSTUVWXYZ

L1

## ladder

A **ladder** is used
to climb up and down.
It is made of wood,
metal, or rope. **Ladders**
look like several H's
joined together
from top to bottom.
Firefighters climb
up and down **ladders.**

**ladder**

## laid

**1.** Hens **lay** eggs.
One hen **laid** an egg
this morning.
**2.** The carpenter
**laid** some boards
on the ground.
**Laid** is a form of **lay.**

## lain

Chris **lies** on his bed.
He has **lain** there
for ten minutes.
**Lain** is a form of **lie.**

## lake

A **lake** is a large
amount of water
that is all in one place.
A **lake** has land
all around it.
**Lakes** are not
as big as oceans.

## lamb

A **lamb** is a young
sheep. **Lambs** like
to eat grass.

## lamp

A **lamp** is used to make light. Most **lamps** use electricity to make light.

## land

1. The **land** is the part of the world that is not water. Mountains, deserts, valleys, fields, meadows, and forests are all part of the **land.**
2. A **land** is a country. Donna has stamps from many different **lands.**
3. The **land** is the earth or ground. The farmers planted potatoes on their **land.**
4. When an airplane **lands,** it comes down to the ground. The airplane **landed** in a field.

## language

**Language** is the way people understand each other when they talk or write. Some people know several **languages.**

## large

**Large** means big. Elephants use a lot of space. They are **large** animals. Mice are not **large.**

## last

1. In the alphabet the letter Z comes after all the others. The **last** letter of the alphabet is Z.
2. Yesterday the ground was dry. Today it is wet. It rained **last** night.

## late

1. Our mail comes in the last part of the morning. It comes **late** in the morning.
2. Nancy got to the baseball game after it started. She was **late** for the game.

## later

Melissa came to school after everyone else. She came to school **later** than the other students in her class.

## laugh

**1.** A **laugh** is a happy sound that people make. There were many **laughs** from the people who watched a funny movie.
**2.** Everybody smiled and made happy noises at the circus clown. The clown made everybody **laugh.** Even the clown **laughed** when he stepped in a bucket of water by mistake.

## laundry

**Laundry** is dirty clothes that are ready to be washed.
**Laundry** is also clean clothes that have just been washed.

laundry

## law

A **law** is a rule that people agree to share. **Laws** are made to make life between people safe and fair.

## lay[1]

**1.** Hens make eggs inside themselves. When hens **lay** eggs, the eggs come out. One hen **laid** an egg this morning.
**2.** The carpenter puts the boards down. He **lays** them down next to his saw and hammer.

lay[1]

## lay²

Chris **lies** on his bed.  He **lay** there because he was tired.
**Lay** is a form of **lie.**

## lazy

A **lazy** person does not want to do any work. Phillip wants to lie on the grass with his eyes closed.  He does not want to cut the grass.  He feels **lazy** today.

## lead¹

1. **Lead** is a kind of metal. It is soft and gray.
**Lead** is very heavy.
2. **Lead** is the part of a pencil that people write with.  Pencil **leads** are made from a special powder that is pressed together.

## lead²

1. Sam shows his friends the way through the forest.  Sam **leads** his friends through the forest.  He **led** them to the lake.
2. The road goes to town.  It **leads** to town.

## leaf

A **leaf** is part of a plant.  It is thin and flat. **Leaves** come in different shapes.  Tree **leaves** are green in the summer.  They become red, orange, or yellow in the fall.

leaf

## learn

The teacher shows Hillary how to spell a word. Hillary **learns** how to spell from her teacher. Yesterday she **learned** how to spell dictionary.

## least

Father has a little cereal. Mother has less cereal. The baby has the **least** cereal. He has less cereal than anybody.

## leather

**Leather** is a kind of animal skin. **Leather** is made into boots, shoes, and gloves.

## leave

**1.** Paula will go away on Thursday. She **leaves** for her vacation then. Paula **left** early on Thursday morning.
**2.** Allison puts her library books on the table. She **leaves** them on the table while she eats dinner.

## leaves

**Leaves** means more than one **leaf. Leaves** change color in the fall.

## led

**1.** Sam **leads** his friends through the forest. He **led** them to the lake.
**2.** The road **led** into town. **Led** is a form of **lead.**

## left[1]

**Left** is the opposite of right. You read these words from **left** to right.

**leather**

**LEFT TURN ONLY**

**left[1]**

## left²

1. Paula **leaves** for her vacation on Thursday. She **left** early on Thursday morning.
2. Allison **left** her books on the table while she ate dinner.

**Left** is a form of **leave.**

## leg

1. The **leg** is a part of the body. People and animals walk on their **legs.** People have two **legs.** Many animals have four **legs.** Spiders have eight **legs.**
2. Each corner of the table has a **leg** under it. The **legs** hold up the top of the table.

## lemon

A **lemon** is a kind of fruit. It is yellow. **Lemons** have a sour taste.

## length

This line is two inches long. _____

Its **length** is two inches. **Lengths** are measured in many different ways.

## less

**Less** is the opposite of more. A quart is not as much as a gallon. A quart of milk is **less** than a gallon of milk.

## let

Polly wants another peach. Her mother says she can have one. Polly's mother **lets** her have another peach. Yesterday she **let** her have two oranges. Polly's mother will not **let** her have too much candy.

## let's

**Let's** is a short way to write **let us.** "**Let's** go to a movie," Connie said to her mother when she came back from school.

## letter

**1.** A **letter** is a symbol used to make written words. It is part of an alphabet. A,B,C,D,E,F,G,H,I,J,K,L,M,N,O,P,Q,R,S,T,U,V,W,X,Y, and Z are all **letters.**

**2.** A **letter** is a message on a piece of paper. Erica wrote a **letter** to her friend.

## lettuce

**Lettuce** is a kind of vegetable. It has large, green leaves.

## library

A **library** is a place where many books are kept. People borrow books from **libraries** to read them at home.

**lettuce**

## lie[1]

**1.** Anyone who **lies** says something that is not true. In a story every time a puppet **lied,** its nose grew.

**2.** In a story a puppet said something that was not true. He told a **lie.** When he told **lies,** his nose grew very long.

## lie[2]

Chris is on his bed. His head is near one end of it. His feet are near the other end. Chris **lies** on his bed. He **lay** there for an hour because he was tired.

## life

**1.** People, animals, and plants are alive. They have **life.**

**2.** A **life** is the amount of time a person, an animal, or a plant is alive. Many insects have short **lives.**

## lift

Rick picks up the top of the pot. He **lifts** it to smell the soup inside the pot. He **lifted** the top later to taste the soup.

## light[1]

**1. Light** is the opposite of dark. Days are **light.** Nights are not.

**2. Light** is what we get from the sun, stars, lamps, and candles.

**3.** Carol starts a fire in the fireplace. She **lights** the fire. She **lighted** a fire because she was cold.

**4.** When the lamp is on, the room is not dark. The lamp **lights** the room.

## light[2]

**Light** is the opposite of heavy. A feather does not weigh much. A feather is **light.**

**lighthouse**

## lighthouse

A **lighthouse** is a kind of building. It is tall and narrow. A strong light turns around at the top of it. **Lighthouses** are built near ocean shores. Their lights warn ships away from rocks at night or in fogs.

## lightning

**Lightning** is a big flash of light. It is a form of electricity. **Lightning** moves from the sky to the ground during storms.

## like[1]

Lucy and Susan are friends. They **like** each other. They **liked** each other the first time they met.

## like[2]

Telephone poles are almost all the same. One telephone pole looks a lot **like** another.

## lime

A **lime** is a kind of fruit. It is green and looks like a small lemon. **Limes** have a sour taste.

## line

1. A **line** is a long, thin mark. **Lines** can be crooked, curved, or straight.
2. Stephanie stood in front of Jim. Jim stood in front of Alex. Alex stood in front of Elizabeth. And Elizabeth stood in front of John. They were in a **line** at the movie theater. Some **lines** at movie theaters are very long.

## lion

A **lion** is a large animal. It looks a lot like a big cat. Male **lions** have thick, long hair around their heads. **Lions** are wild animals. They like to eat meat.

lion

## lioness

A **lioness** is a female lion. **Lionesses** do not have thick, long hair around their heads.

## lip

A **lip** is part of the face. It is around the edge of the mouth. People have two **lips.**

## liquid

A **liquid** is something that pours. A **liquid** in a bottle will take the shape of the bottle. Water, milk, and juice are **liquids.**

## list

Jackie wrote the names of her favorite foods on a piece of paper. Together the names made a **list.** Jackie also made many **lists** of her favorite books and flowers.

## listen

Greg can hear his teacher speak. He **listens** to the words she says. Greg **listened** to his teacher until school was over.

## liter

A **liter** is an amount of a liquid. Quarts are almost as big as **liters.**

## little

**Little** means small. Babies are **little** people.

## live

1. People, animals, and plants **live** while they grow and change. The flowers **lived** during the summer.
2. Peter has his home in the city. He **lives** in the city.

## lives

**Lives** means more than one **life.** Many plants have short **lives.**

## load

**1.** A **load** is something to be carried. Phil took two **loads** of books to the library.

**2.** Liz and her mother put the furniture onto the truck. They **load** the truck. They **loaded** boxes onto the truck too.

## loaf

A **loaf** of bread is an amount of bread. It is baked in one piece. Supermarkets sell many **loaves** of bread to people.

## loan

**1.** Pete forgot his pencil. Carla will let him use one of her pencils. She will **loan** him a pencil. Last week she **loaned** him a pen.

**2.** Pete uses Carla's pencil. It is not his pencil. It is a **loan** from Carla. Carla has extra pencils for **loans** to the other students.

loaf

## loaves

**Loaves** means more than one **loaf.** Supermarkets sell many **loaves** of bread.

lobster

## lobster

A **lobster** is an animal. It lives in the ocean. **Lobsters** have hard shells and big claws.

## lock

1. A **lock** is something used to keep an object shut. **Locks** are opened and closed with keys. Many doors have **locks** on them.

2. Arthur turns the key in the **lock**. He **locks** the door with the key. He **locked** all the doors to his house before he went on a vacation.

lock

## log

A **log** is a piece of wood. **Logs** are cut from trees. Some **logs** are burned in fireplaces. Other **logs** are cut into boards.

## long

1. **Long** is the opposite of short. Swans have **long** necks.

2. The soccer game will take an hour to play. The game will be an hour **long**.

log

## look

People see monkeys at the zoo. They **look** at the monkeys. They **looked** at other animals too.

## loose

**Loose** is the opposite of tight. Mary tries to walk in her mother's shoes. But they are too big for her. The shoes are **loose** on her feet. Her feet come out of them when she tries to walk.

## lose

1. Becky does not know where her gloves are. She can't find them. She **loses** a lot of gloves. Becky **lost** three pairs of gloves last winter.
2. Jim and Mark play a game together. One of them will win the game. The other one will **lose** the game.

## lost

1. Becky **loses** gloves a lot. She **lost** three pairs of gloves last winter.
2. Mark **lost** the game he played with Jim.
3. Jill stood in the forest. She did not know how to get out of it. She was **lost** in the forest.
**Lost** is a form of **lose.**

## lot

Many stars shine in the night sky. There are a **lot** of stars to watch.

## loud

Thunder is heard for many miles in a storm. It makes a lot of noise. Thunder is **loud.**

## love

Nick feels good when he is with his family. He wants only nice things to happen to his family. Nick **loves** his family. He always **loved** his family.

## low

**Low** is the opposite of high.
1. Some things are not very high. They are **low.** Jim sat on a **low** chair to eat his lunch in the kitchen.
2. Some clouds are near the ground. **Low** clouds make fog.
3. Vegetables do not cost much in the summer. They have a **low** price.

## luck

Megan hoped the weather would be nice
for the class picnic. If her **luck** was good, the
sun would shine. If her **luck** was bad, everyone
would need umbrellas.

## lumber

**Lumber** is wood that has been cut into boards.
**Lumber** comes from very big trees.

## lumberjack

A **lumberjack** is a person whose job is to cut
down very big trees. **Lumberjacks** use big saws
and axes. They make the trees into logs. Then
they bring the logs
to a place
where they are cut
into boards.

lumberjack

## lunch

**Lunch** is a meal. It is eaten in the middle
of the day. People who eat small breakfasts
often eat big **lunches.**

ABCDEFGHIJKL**M**
NOP**Q**RSTUVWXYZ

**Mm**

## machine

A **machine** is an invention. It does work for people. It is made of metal, wood, glass, rubber, or other things. Cars, trains, and refrigerators are all **machines.**

## mad

**Mad** means angry. Bill missed the bus to school. He is **mad** at himself because now he must walk to school.

## made

1. Ellen **makes** breakfast every morning. She **made** breakfast for her father on his birthday.
2. The happy song **made** Marjorie feel good.
3. Sally **made** a nickel for every newspaper she sold.
4. Frank **made** his bed only on Saturday.
**Made** is a form of **make.**

**machine**

## magic

1. **Magic** is a kind of power used in many stories. In one fairy tale **magic** made a princess sleep for a hundred years.
2. Some people can do tricks that seem impossible. These are **magic** tricks.

## magician

1. A **magician** is a person in a story who does magic. Some **magicians** can fly. Others can change people into animals.
2. A **magician** is a person who does tricks in a show.

## magnet

A **magnet** is a piece of metal or rock that will stick to iron. **Magnets** are used in many machines.

## mail

**1.** The **mail** is a system we use to send letters and packages from one place to another.
**2.** Alex's parents send him packages at camp. They **mail** a package to him every week. The last package they **mailed** to him had cookies and books in it.

**magnet**

## make

**1.** Ellen likes to cook breakfast. She **makes** breakfast every morning. She **made** breakfast for her father on his birthday.
**2.** Music can change Marjorie's mood. Happy songs **make** her feel good.
**3.** Sally gets a nickel for each newspaper she sells. She **makes** money then.
**4.** Sometimes Frank folds the edges of his blanket under his mattress. Then the blanket is smooth and neat on top of his bed. Frank **makes** his bed when he remembers to do it.

## male

There are **male** and female people and animals in the world. Boys and men are **male** people.

**mail**

## man

A boy grows up to be a **man**. Boys grow up to be **men**.

## manage

Bill leads our baseball team. He tells us what to do. He **manages** the team so that we will win games. Bill **managed** the team last year too.

## many

**Many** is the opposite of few. The beach is not covered with one or two grains of white sand. The beach is covered with **many** grains of white sand.

map

## map

A **map** is a picture that shows where places are. **Maps** of countries show cities, towns, rivers, lakes, mountains, and deserts.

## maple

A **maple** is a kind of tree. **Maple** trees are used to make furniture and syrup.

## marble

1. **Marble** is a kind of stone. **Marble** is a part of some buildings. Artists use **marble** to make statues. Most **marble** is white. Other kinds of **marble** have colors mixed in.
2. A **marble** is a small glass ball. **Marbles** are used in several games.

maple

## march

People in the army walk with the same size steps when they walk together. They **march** together a lot. They **marched** across the field ten times before they took a rest.

## March

**March** is a month of the year. It has 31 days. **March** comes after February and before April.

## mark

**1.** Susan walked on the kitchen floor with dirty shoes. One shoe made a **mark** of dirt near the refrigerator. Her shoes made other **marks** near the table.
**2.** The teacher puts a check on a test. She **marks** the right answers with checks. She **marked** several tests with checks. Then she gave the tests to the students.

## marmalade

**Marmalade** is a kind of food. It is soft and sweet. **Marmalades** are made from fruits.

## marry

A man and a woman often agree to share their lives together. They **marry** each other by the laws of where they live. After Tony **married** Rose, he became her husband. Rose became Tony's wife.

## marsh

A **marsh** is a piece of land. It is soft and wet. Frogs and mosquitoes live in **marshes.**

## marshmallow

**Marshmallow** is a kind of candy. It is soft and white. **Marshmallows** are pieces of this candy. They are good to eat.

## mask

A **mask** is something to cover the face. **Masks** can be made of paper, cloth, plastic, or rubber. Dennis wore a red and green **mask** with his Halloween costume. He then went to see his friends.

## match

The color of Janet's shirt and shoes are the same. They **match** each other. She bought the shoes because they **matched** the shirt.

## mattress

A **mattress** is the top part of a bed. It is soft. People sleep on **mattresses.**

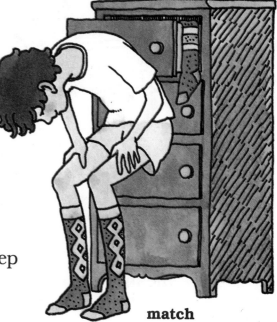

**match**

## may

1. There is a chance it will snow today. It **may** snow. It **might** have snowed yesterday, but it didn't.
2. Larry wants to go to the store. He **may** go when his mother says so.

## May

**May** is a month of the year. It has 31 days. **May** comes after April and before June.

## maybe

There may be a party at Jamie's house tomorrow. **Maybe** Jamie will have a party tomorrow.

## me

I bought a kite at the store. After I bought it, the kite belonged to **me.**

## meadow

A **meadow** is an area of ground. **Meadows** are covered with long grass.

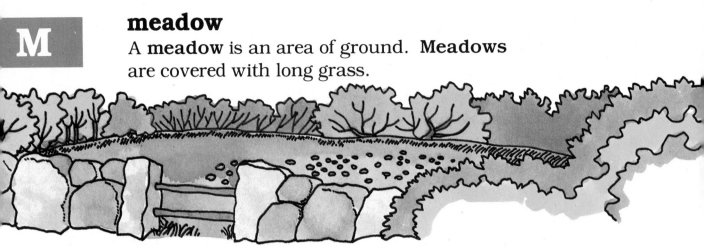

meadow

## meal

A **meal** is an amount of food that people eat at one time. Breakfast and lunch are **meals.**

## mean[1]

1. Large is another word for big. Large and big **mean** the same thing. They never **meant** anything else.
2. Allison wants to play the violin well. It is important to her. It **means** a lot to her.

## mean[2]

**Mean** is the opposite of nice. **Mean** people are not kind or friendly.

## meant

1. Large and big **mean** the same thing. They never **meant** anything else.
2. It **meant** a lot to Allison to play the violin well.

**Meant** is a form of **mean.**

## measure

Ben finds out how wide his house is. He counts the number of feet along the front. He **measures** the width of his house. Then he **measured** its length.

## meat

**Meat** is a kind of food. It comes from animals. Chicken, beef, and ham are all **meats.**

## medicine

Sick people use **medicine** to help themselves get well. Many **medicines** are made from plants. Some of them are made to eat or drink.

## meet

**1.** On her first day at school Carol gets to know some of the other students. She **meets** them in her class. She **met** six students the first day.
**2.** Ann and Barbara plan to have lunch together at noon. They will go to the restaurant at the same time. They will **meet** there at noon.

## melt

When ice gets warm, it **melts** and becomes water. The ice on the lake **melted** in the spring.

## memory

Wendy remembers everything that happened on her vacation. She has a good **memory.** Her brothers don't remember much about it. Their **memories** are not very good.

## men

**Men** means more than one **man.** Boys grow up to be **men.**

## mess

The clothes in Leslie's room were all over the bed. Books and papers covered the floor. Leslie's room was a **mess. Messes** are not neat.

## message

A **message** is a group of words that is sent between people. Many **messages** are sent through the mail.

## met

**1.** On her first day at school Carol **meets** the other students in her class. She **met** six of them the first day.

**2.** Ann and Barbara **met** for lunch at a restaurant.

**Met** is a form of **meet**.

## metal

Iron, gold, silver, copper, and lead are all kinds of **metal**. **Metal** is usually found underground. It is strong and hard. It can bend and not break. **Metals** can be melted and shaped into many things.

## meter

A **meter** is an amount of length. A yard is almost as long as a **meter**. One kilometer is 1,000 **meters**.

## mice

**Mice** means more than one **mouse**. **Mice** do not like cats.

## microscope

A **microscope** is used to see things we cannot see well with our eyes alone. **Microscopes** can make tiny things look big. They can make a grain of sand look as big as a penny.

## midday

**Midday** is the middle of the day. People eat lunch at **midday**.

**microscope**

## middle

An elbow is in the **middle** of the arm. It is not close to either end.

## might

**1.** It **may** snow today. It **might** have snowed yesterday, but it didn't.
**2.** Larry **might** have gone to the store, if his mother had said he could.
**Might** is a form of **may.**

## mile

A **mile** is an amount of length. It is 5,280 feet long. Three **miles** is about the same length as five kilometers.

## milk

**Milk** is a kind of liquid. It comes from cows, goats, and other animals. People drink **milk.** Parts of it are made into butter and cheese.

## million

A **million** is a number. One **million** is written 1,000,000 . It takes a thousand thousands to make one **million.** The sun is **millions** of miles away from us.

## mind

The **mind** is part of a person that thinks, feels, learns, remembers, wishes, and imagines. Without **minds** humans would not be humans.

## mine

This book belongs to me. This book is **mine.**

## minus

If you take two away from six, you get four. Six **minus** two is four. Six **minus** two can also be written 6−2.

## minute

A **minute** is a short amount of time. There are 60 seconds in one **minute.** There are 60 **minutes** in one hour.

## mirror

A **mirror** is a piece of glass you can see yourself in. Clothes stores have many **mirrors** in them.

## miss

**1.** Peter throws the basketball at the basket. The ball does not go through the basket. The ball **misses** the basket. The basketball **missed** the basket the next time Peter threw it too.
**2.** Sarah is cold in the winter. She wishes she could be warm. She **misses** summer when she is cold.

**mirror**

## Miss

**Miss** is a word people use with a woman's name if she has not married. **Miss** Brown owns the store on our street.

## mistake

Kim did not bring an umbrella to school. She did not think it would rain. It rained. Kim had made a **mistake.** It is easy to make **mistakes** about the weather.

## mix

Tom puts eggs and flour in a bowl. Then he pushes them into each other with a big spoon. He **mixes** them together. Tom **mixed** other things with the eggs and flour. Then he baked a cake.

## model

A **model** is a small copy of a large object. Mark built three **models** of planes. Each **model** was a foot long.

**model**

## moment

A **moment** is a very small amount of time. The light in the lighthouse flashed for just a **moment.**

## Monday

**Monday** is a day of the week. **Mondays** come after Sundays and before Tuesdays.

## money

People use **money** to buy things. They are paid **money** for the work they do. Most **money** is made of paper or metal. Pennies, nickels, dimes, quarters, and dollars are all kinds of **money.**

## monkey

A **monkey** is a kind of animal. It has long arms and legs. It lives in a tree. **Monkeys** look more like people than most animals.

## monster

A **monster** is a huge, evil animal or person. The dragons in most stories are **monsters.**

**money**

**monster**

## month

A **month** is one of the 12 parts of a year. The **months** in order are January, February, March, April, May, June, July, August, September, October, November, and December.

## mood

Jack was happy because it was sunny outside. He was in a good **mood.** Then it rained. Now Jack is in a bad **mood.** His **moods** change as often as the weather.

## moon

The **moon** is a bright, big object in the sky at night. It may look like a big silver ball or like just a piece of one.

**moon**

## more

1. **More** is the opposite of less. There are six children in the blue car. There are 30 children in the yellow bus. There are **more** children in the yellow bus than in the blue car.
2. Ned ate a turkey sandwich. Then he ate another one. Ned ate **more** food because he was hungry.

## morning

**Morning** is part of the day. The sun rises in the **morning. Mornings** end at noon.

## mosquito

A **mosquito** is a kind of insect. It has wings. Female **mosquitoes** bite people and animals. Nobody likes them.

mosquito

## most

The sun shines with more light than anything else we see. It shines with the **most** light of anything we see.

## moth

A **moth** is an insect. **Moths** look like small butterflies, but their wings are not as pretty. A caterpillar changes into a **moth** in a cocoon.

moth

## mother

A **mother** is a woman who has at least one child. **Mothers** and fathers take care of their children.

## motor

A **motor** is a machine. It uses energy from electricity or gas. Refrigerators use **motors** to keep food cold. Cars use **motors** to make them move.

## motorcycle

A **motorcycle** is a machine. It is like a big, heavy bicycle with a motor. Some **motorcycles** can go as fast as cars.

motorcycle

## mountain

A **mountain** is an area of land. It rises high above the land around it. The tops of some **mountains** are always covered with snow.

## mouse

A **mouse** is a small animal. It has big ears, short fur, and a long tail. **Mice** do not like cats.

## mouth

The **mouth** is a part of the head. It is under the nose. People and animals eat food with their **mouths.**

**mouse**

## move

**1.** George goes inside when it starts to rain. He **moves** inside to stay dry. He **moved** into the house quickly.
**2.** Laura lives in an apartment now. She will live in a house soon. She will **move** to a house this spring.

## movie

A **movie** is a story made with pictures. The pictures are made with a special camera. People watch **movies** in theaters or on television.

## Mr.

**Mr.** is a word people use with a man's name. **Mr.** Carpenter teaches first grade.

## Mrs.

**Mrs.** is a word people use with a woman's name if she is married. **Mrs.** Smith took her children to the circus.

## Ms.

**Ms.** is a word people use with a woman's name.
**Ms.** Cook drives a bus for the city.

## much

**1.** David wondered about the price
of the basketball he saw. "How **much** does
that basketball cost?" he asked.
**2.** Jessica wanted a camera for her birthday.
She wanted it more than anything else. She
wanted it very **much.**

## mud

**Mud** is very wet dirt. Pigs like to roll in the **mud.**

## multiply

When you **multiply** two numbers together, you
add one number to itself several times. The
symbol for **multiply** is ×. 2 × 3 is the
same as 2 + 2 + 2. Any number
**multiplied** by 2 is an even number.

## muscle

A **muscle** is a part of the body. It is
under the skin. People and animals
use **muscles** to move.

## museum

A **museum** is a place
that collects, keeps, and shows
things. It can have art,
machines, or models
of animals in it. People go
to **museums** to see things
they cannot own.

## mushroom

A **mushroom** is a kind of
plant. It grows in dark places.
**Mushrooms** look like small umbrellas.

**mushroom**

## music

Music is a pretty sound.  People sing and play instruments to make **music.**

## musical

A bell makes a pretty sound.  It is a **musical** sound.

## must

The sun does not decide to rise each morning.  It has no choice.  The sun **must** rise every day.

## mustard

**Mustard** is a yellow food.  It has a strong, hot taste.  People eat **mustard** on hot dogs or in sandwiches.

## my

That bicycle belongs to me.  It is **my** bicycle.

## myself

I see **myself** when I look in a mirror.

## mystery

A **mystery** is something that a person does not understand.  Pam does not understand what makes plants grow.  She wonders about it.  It is a **mystery** to her.  Nature is full of **mysteries.**

A B C D E F G H I J K L M
N O P Q R S T U V W X Y Z

# Nn Oo

## nail

**1.** A **nail** is a piece of metal. It is short and thin. It has a point at one end. The carpenter hits **nails** into a board with a hammer.

**2.** A **nail** is a part of the body. It is the hard part at the end of the fingers and toes. Sally is often told not to bite her **nails.**

## name

**1.** A **name** is a word that people use to call something by. All things have **names.**

**2.** Polly is a **name** for a girl or a woman. Greg is a **name** for a boy or a man.

## narrow

**Narrow** is the opposite of wide. The two houses are close together. The yard between the houses is **narrow.**

## naturally

Things that happen **naturally** are things that people don't make happen. The seasons come **naturally** each year.

## nature

People, animals, plants, the sky, the land, the ocean, and the weather are all parts of **nature.** Only things that people make are not part of it.

**nail**

## near

**Near** is the opposite of far. The hairs on our heads grow close together. They grow very **near** each other.

## neat

The floor and furniture in Linda's room are clean. Her clothes are in her closet. Her books are on her desk. Her bed is made. Linda's room is **neat.**

## neck

The **neck** is a part of the body. It joins the head to the shoulders. Giraffes have long **necks.**

## need

Plants must have water to live. They **need** water to live. When no rain fell for weeks, the plants **needed** a lot of water.

## needle

needle

1. A **needle** is a kind of tool. It is a short and very thin piece of metal. **Needles** have holes at one end and sharp points at the other. People use **needles** and thread to sew clothes.
2. A **needle** is a kind of leaf. It is shaped like the **needle** people use to sew with. This **needle** is found on pine trees.

## neighbor

Paul and Joe live near each other. Paul is Joe's **neighbor.** Joe is Paul's **neighbor.** Many **neighbors** are friends.

## neighborhood

A **neighborhood** is an area of a city or a town. Neighbors live in the same **neighborhood.** Children from several **neighborhoods** often go to one school.

## neither

**Neither** means not either. After Joel planted seeds in his garden, his hands were dirty. **Neither** hand was clean.

## nest

A **nest** is the home of some birds. It is made to hold eggs. Birds make **nests** from grass, mud, sticks, string, and other things.

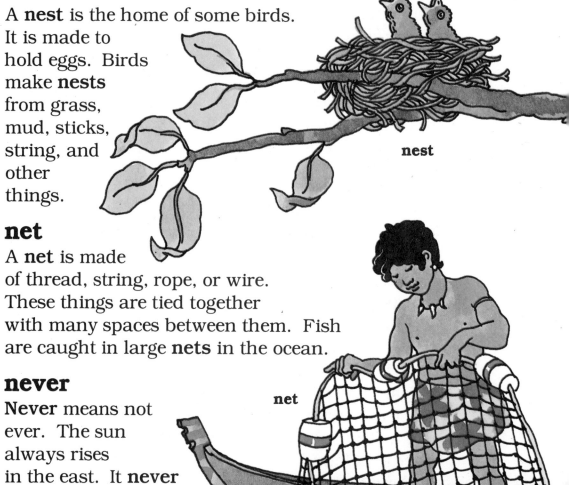

nest

## net

A **net** is made of thread, string, rope, or wire. These things are tied together with many spaces between them. Fish are caught in large **nets** in the ocean.

net

## never

**Never** means not ever. The sun always rises in the east. It **never** rises in the west.

## new

New is the opposite of old. Eliza bought a bicycle. It had just been made. It was a **new** bicycle.

## news

News is a story about things that happen. Jeff reads the **news** in the newspaper.

## newspaper

People write about news in a **newspaper.** It has large pages. **Newspapers** have pictures and stories in them.

## next

This summer Edward will stay home. The summer after this one he will go to camp. **Next** summer he will go to camp.

## nice

Nice is the opposite of mean. **Nice** people are friendly and kind.

## nickel

A **nickel** is a kind of coin. It will buy the same things as five pennies. Five **nickels** will buy the same things as one quarter.

## night

Night is the opposite of day. **Night** begins when the sun sets. It ends when the sun rises. People see the moon and the stars at **night**. **Nights** are dark.

## nightmare

A **nightmare** is a bad dream. Karen had **nightmares** after she saw the movie.

## nine

Nine is a number. **Nine** is written 9. $8 + 1$ is 9.

## no

**1. No** is the opposite of yes. Maureen asked her mother if she could have another dessert. Her mother said, **"No."**

**2.** Peter is sick. He was sick yesterday too. He feels **no** better today than he did yesterday.

**3. No** means not any. The cereal box is empty. There is **no** cereal in it.

## nobody

**Nobody** means no person. **Nobody** can fly like a bird.

## nod

When people move their head up and down, they **nod.** Larry **nodded** when he agreed with Nancy.

## noise

A **noise** is a sound. Fireworks make loud **noises.**

## noon

**Noon** is a name for 12 o'clock. It is the middle of the day. People eat lunch at **noon.**

## north

North is a direction. **North** is the opposite of south. If you look where the sun rises, **north** is on your left.

**north**

## nose

The **nose** is part of the face. It is below the eyes and above the mouth. People and animals smell with their **noses.**

## not

The sun shines on us during the day. It does **not** shine on us at night.

## note

**1.** A **note** is a short sound of music.
It can be written as a symbol.
Music is made of **notes.**
**2.** A **note** is a short, written
message.

## nothing

**Nothing** means no thing. An empty bottle
has **nothing** in it.

## notice

Jill sees that the leaves have begun
to change color. After she **notices** this,
she knows autumn has begun. Jill also
**noticed** that the days were not as warm
as they had been.

## November

**November** is a month of the year. It has 30
days. **November** comes after October and
before December.

**note**

## now

Anyone who reads these words, reads them
at this moment. You read these words **now.**

## number

**1.** A **number** is a symbol for an amount
of things. People use **numbers** to count things.
**2.** Kelly uses a **number** to call her friends
on the telephone. Her **number** is 555-1234.

## nurse

A **nurse** is a person whose job is to help take
care of sick people. **Nurses** work with doctors.

## nut

A **nut** is a seed or a dry fruit. It has a hard
shell. Many **nuts** grow on trees.

## oak

An **oak** is a kind of tree. **Oak** trees are big. Acorns grow on **oaks**.

## obey

Larry does what he is told by his parents. He **obeys** them. Yesterday Larry **obeyed** them when he arrived early for dinner.

## object

An **object** is anything that is not alive that people can see or touch. Buildings, tables, books, and scissors are all **objects.**

oak

## ocean

An **ocean** is a very large amount of water. The water has a lot of salt in it. **Oceans** cover almost three quarters of the world.

## o'clock

**O'clock** is a word used to say what time it is. Seven **o'clock** in the morning is the time when I get up.

## October

**October** is a month of the year. It has 31 days. **October** comes after September and before November.

## odd

1. **Odd** is the opposite of even. Some **odd** numbers are 1,3,5,7, and 9.
2. **Odd** means strange. The smell in the kitchen was a mystery to Jeff. It was an **odd** smell.

## of

**1.** The table is made from wood. It is a table **of** wood.

**2.** Paul carried a bag filled with food. He carried the bag **of** food into his house.

**3.** It is two minutes before five o'clock. It is two minutes **of** five.

## off

**1.** **Off** is the opposite of on. The house is dark at night when the lamps are **off.**

**2.** Peter picked up a book from the table. He took the book **off** the table.

## offer

Janet wants to help her parents clean their yard. She **offers** to rake the leaves. She **offered** to rake them on Saturday.

## office

An **office** is a place where people work. Some buildings have many **offices.**

## often

Anything that happens again and again happens **often.** It **often** rains in April.

## oil

Oil is a kind of liquid. It floats on water. **Oils** from animals and vegetables are used in foods. Oil from rocks is used to make cars run.

## old

**1.** **Old** is the opposite of new. The **old** pencil was just two inches long. It had been used many times.

**2.** **Old** is the opposite of young. Some grandfathers and grandmothers are **old** people.

**3.** Jennifer was born three years ago. She is three years **old.**

**O**

## on

**1.** **On** is the opposite of off.  The room is bright when the lamp is **on.**
**2.** Mary moves the dishes from the sink to the table.  She puts them **on** the table.
**3.** The cow took a bite of grass.  It chewed **on** the grass for a long time.
**4.** George has a book about dragons. The book is **on** dragons.  It tells all about them.
**5.** Summer vacation begins next Saturday.  It starts **on** Saturday.

## once

**1.** The parade went through town only one time.  It went past the crowds only **once.**
**2.** **Once** means after.  We will play outside **once** the rain stops.

## one

**1.** **One** is a number. **One** is written **1.**
**2.** Sarah will eat a green grape but not a purple **one.**  Purple grapes have seeds in them.  She does not like the **ones** with seeds.

## onion

An **onion** is a kind of vegetable.  It is round. It has a strong smell. **Onions** grow underground.

onion

## only

There is one and no more than one sun in the sky.  There is **only** one sun.

## open

**Open** is the opposite of close.

**1.** People can enter the supermarket at eight o'clock in the morning. The supermarket **opens** at eight o'clock. It **opened** an hour late after a winter storm.

**2.** Maggie takes the top off a box. She **opens** the box to see what is inside it.

## opposite

**1.** Day is the **opposite** of night. Two things that are completely different from each other are **opposites.**

**2.** The directions north and south are at either end of the same line. The two directions are **opposite** each other.

## or

It could rain today. It could snow today. It could rain **or** snow today.

## orange

**1.** An **orange** is a kind of round fruit. It is about the size of a tennis ball. **Oranges** grow on trees.

**2.** Orange is a color. Pumpkins are **orange**.

## orchestra

An **orchestra** is a large group of people who play instruments together. Many people pay money to hear **orchestras** play music.

**orchestra**

## order

The letters in the alphabet always follow each other in the same way. A,B,C,D,E,F,G,H,I,J, K,L,M,N,O,P,Q,R,S,T,U,V,W,X,Y, and Z is the **order** of the alphabet.

## ostrich

An **ostrich** is a huge bird. It has long legs and a long neck. **Ostriches** cannot fly, but they can run very fast.

ostrich

## other

1. Vicki wears two socks on her feet. One sock looks fine. The **other** sock has a hole in it.

2. Kim has no time to play today. She will have time to play some **other** day.

3. The barber shop is not on this side of the street. It is on the **other** side.

## ounce

An **ounce** is an amount of weight. One **ounce** is about the same as 28 grams. 16 **ounces** is one pound.

## our

These clothes belong to us. They are **our** clothes.

## ours

Karen is my friend. She is also my sister's friend. Karen is a friend of **ours.**

## ourselves

Alan and I looked in a mirror. We saw **ourselves** there.

## out

1. **Out** is the opposite of in. The wind blew in one window and **out** another.
2. Some clouds hid the moon behind them. When the clouds moved, the moon came **out** again.
3. Joe pushed the bottle from the window. The bottle fell **out** the window and broke.

## outside

**Outside** is the opposite of inside.
1. Donna goes out of her house. She goes **outside** to play in the snow.
2. The shell is the **outside** of the egg. It is very different from the inside of the egg.

## oval

An **oval** is a shape. It looks like a circle that somebody sat on. Eggs are shaped like **ovals.**

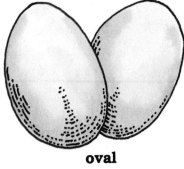

**oval**

## oven

An **oven** is a metal box that food is cooked in. A restaurant often has many **ovens** in its kitchen.

## over

**1.** The plane flies above the ocean. It flies **over** the water.
**2.** Jackie wore a sweater on top of her shirt. She wore the sweater **over** her shirt.
**3.** An elephant is more than seven feet tall. It is **over** seven feet tall.
**4.** The baby turned her glass of milk **over.** The milk poured all **over** the floor.
**5.** The band played the song again. They played it **over** because everybody liked it.
**6.** The movie has ended. It is **over.**

oven

## owe

Beth ate an apple in a store. She **owes** the store money for the apple. She **owed** the money until she paid for the apple.

owl

## owl

An **owl** is a bird. It has a large head and large eyes. **Owls** sleep during the day.

## own

Jane buys a book. It is now hers. She **owns** the book. Jane **owned** a lot of books.

ABCDEFGHIJKLM
NOP QRSTUVWXYZ

# Pp

## pack

Walter puts clothes in a trunk. He **packs** his clothes in the trunk. Walter **packed** his clothes for camp.

## package

A **package** is a kind of box. It is covered with paper. Letters and **packages** are sent through the mail.

## page

A **page** is one side of a piece of paper. These words are written on a **page**. Books are made of many **pages** held together.

## paid

**1.** Paula **pays** for some fruit at the store. She **paid** for the fruit with money.
**2.** Phillip **paid** attention to everything his teacher said to the class. **Paid** is a form of **pay**.

package

## pail

A **pail** is used to hold things. It is made of metal or plastic and has a flat, round bottom. **Pails** are the same shape and size as buckets.

pail

## pain

Jerry cuts his finger with a knife. His finger hurts. Jerry has a **pain** in his finger. Jerry has **pains** in his teeth.

## paint

**1. Paint** is a kind of liquid with color in it. People cover things with **paint.** Artists use **paints** to make pictures too.

**2.** Tom covers the walls of his room with **paint.** He **paints** his room. He **painted** the walls all day.

## pair

**1.** Two shoes that match each other are a **pair** of shoes. Alexander has three **pairs** of shoes.

**2.** A **pair** of scissors is like two knives joined together. Some **pairs** of scissors are used to cut paper.

## pajamas

**Pajamas** are a kind of clothes. People wear them while they sleep. **Pajamas** are soft and warm.

## palace

A **palace** is a kind of building. It is a large and beautiful home. Kings and queens live in **palaces.** Some **palaces** have hundreds of rooms.

paint

palace

## palm

The **palm** is the inside part of the hand. There are many lines on the **palms** of our hands.

## pan

A **pan** is something to cook in. It is made of metal. **Pans** are wide and not very deep.

## pancake

A **pancake** is a kind of food. It is a thin, flat cake. **Pancakes** are made of flour, eggs, and milk mixed together. They are cooked in hot butter.

palm

## pants

**Pants** are a kind of clothes. People wear **pants** over their legs. Many pairs of **pants** have pockets on the sides and near the top.

## paper

**Paper** is something people use to write on. It is made from wood. Books, magazines, and newspapers are all made of **paper.** Colored **papers** are often used to wrap things.

## parade

A **parade** is a group of people, bands, or cars that follow each other on a street. Cities and towns have **parades** on holidays.

parade

park

## parent

A **parent** is the mother or the father
of a child. Men and women are **parents.**

## park

1. A **park** is an area of land. It is covered
by grass and trees. People in cities enjoy **parks**
because they do not have land of their own.
2. Ned's father puts his car into a garage. He
**parks** the car there. He **parked** the car
in the garage every night during the winter.

parrot

## parrot

A **parrot** is a bird. It has a large beak. **Parrots**
have feathers in several bright colors. Some
**parrots** can be taught to say a few words.

## part

One branch is not a whole tree. It is **part**
of a tree. Leaves, roots, and trunks
are other **parts** of trees.

## party

When people get together to have fun, they have a **party**. Many children have **parties** on their birthdays.

## pass

**1.** The wind blows through the forest. It **passes** through the trees. The wind shook leaves as it **passed.**

**2.** At supper George takes corn from a bowl. Then he gives the bowl to Brian. George **passes** the bowl to Brian across the table.

**3.** Michael does well on tests. He **passes** all the tests he takes.

## past

**1.** The **past** is a part of time. It is the part that has already happened. Yesterday is part of the **past.**

**2.** The river goes by many places. It goes **past** several towns.

## patch

**1. Patch** means area. Many berries grow in the meadow. The meadow is a good berry **patch.**

**2.** A **patch** is a small piece of cloth. A scarecrow has **patches** sewn over the holes in its clothes.

## path

A **path** is a kind of trail. **Paths** are made where many people have walked.

**path**

## pay

**1.** Paula buys fruit at the store. She **pays** for it with money. She **paid** a lot for a basket of berries.
**2.** Phillip listens carefully to his teacher. He **pays** attention to everything his teacher says to the class.

## pea

A **pea** is a kind of vegetable. It is small, round, and green. **Peas** are the seeds of a plant.

## peach

A **peach** is a kind of fruit. It is mostly yellow and red. **Peaches** are about the same size as apples. They grow on trees.

## peanut

A **peanut** is a kind of food. It is the seed of a plant. It looks and tastes like a nut. **Peanuts** grow in shells.

## peanut butter

**Peanut butter** is a kind of food. It is made from peanuts. It is soft and smooth. Many people eat **peanut butter** sandwiches.

## pear

A **pear** is a kind of fruit. It is yellow and brown. **Pears** are round at both ends, but their tops are not as big as their bottoms. **Pears** grow on trees.

pea

peach

peanut

## pen

A **pen** is something to write with. It is usually made of metal or plastic. **Pens** use ink.

## pencil

A **pencil** is something to write with. **Pencils** are made of wood. The part that writes on paper is called lead.

## penguin

A **penguin** is a bird. It lives in very cold places near the ocean. **Penguins** have narrow wings. They cannot fly, but they use their wings to swim.

penguin

## penny

A **penny** is a kind of coin. It is a very small amount of money. One **penny** is one cent. 100 **pennies** will buy the same things as one dollar.

## people

Men, women, and children are all **people.**

## pepper

People put **pepper** on food. It can come as tiny black and white grains, or it can be made into a fine powder. **Pepper** has a hot flavor.

## perfect

A carpenter built a house for her dog. She made no mistakes. There was nothing wrong with the house. It was a **perfect** doghouse.

## perhaps

Maybe it will rain today. Maybe it won't.
**Perhaps** it will rain.

## person

A **person** is a girl, a boy, a woman, or a man.
All **persons** are people.

## pet

Any animal that lives with people is a **pet**.
Many people have dogs, cats, or hamsters as
**pets.**

## phone

**Phone** is another name for **telephone.** It is a
machine. A **phone** sends voices
from one place to another. **Phones** use
a small amount of electricity. A **phone**
rings when someone calls.

**phonograph**

## phonograph

A **phonograph** is a machine
that plays records.
**Phonographs**
use electricity
to make them
work.

## piano

A **piano** is
a large
instrument. It
has 88 black
and white keys.
People touch the
keys to make
music. **Pianos**
come in different
shapes and sizes.

**piano**

## pick

Bonnie takes the apples off the tree. She **picks** only the red apples. She **picked** enough red apples to fill a basket.

## picnic

A **picnic** is a meal that people eat outside. Ants like **picnics** as much as people do.

## picture

**1.** Karen draws a lighthouse on a piece of paper. She makes a **picture** of a lighthouse. Many **pictures** are made with pencils, crayons, pens, or paints.

**2.** Diane used her camera on her trip. She used her camera to get a **picture** of the castle. The castle in the **picture** looks just like the real castle. Diane also likes to take **pictures** of people and animals.

## pie

A **pie** is a kind of food. It is round. Its bottom and sides are hard and the inside is soft. The bottoms and sides are made of flour, milk, butter, or other things. Many **pies** are filled with fruit or cheese. Some **pies** have tops that are hard. **Pies** are baked in ovens.

pie

## piece

David cut an orange into four parts. He, Kathy, John, and Terry each had one **piece** of it. There were no more **pieces** of the orange left.

## pig

A **pig** is an animal. It is short and round. It has a flat nose. Many people eat meat that comes from **pigs.**

## pile

There are a lot of clothes on Nora's bed. Everything is on top of everything else. It is a big **pile** of clothes. There are two small **piles** of clothes on her chair. Nora has a lot to put away.

## pilot

A **pilot** is a person who flies planes, jets, or helicopters. Some **pilots** fly several times each week.

## pin

A **pin** is a piece of metal. It is short, thin, and straight. It has a sharp point. **Pins** are used to hold clothes together while they are sewn.

## pine

A **pine** is a kind of tree. **Pines** have needles on their branches. The needles stay green all the time.

## pink

**Pink** is a color. Jill has played in the snow for an hour. Her cheeks are **pink.**

## pint

A **pint** is an amount of liquid. Two **pints** are the same as one quart.

pig

pine

## pipe

A **pipe** is a large tube. It is made of metal, glass, or clay. Many **pipes** are used to carry liquids from one place to another.

pipe

## pirate

A **pirate** is a person who robs the people on ships. Hundreds of years ago **pirates** sailed all over the world.

## pizza

**Pizza** is a kind of food. It is flat and usually round. **Pizzas** are baked with cheese, vegetables, or meat on top of bread.

## place

A **place** is somewhere for a person or thing to be. Rooms, fields, and countries are all **places**. Even a cheek on a face is a **place**.

## plan

1. A **plan** is a group of ideas about what to do. Some **plans** work better than others.
2. Bob has a ticket for the football game. He expects to go to the game. He **plans** to be there. Bob **planned** to go to the game in a car.

## plane

**Plane** is another name for **airplane**. It has two wings and engines that make it fly through the air. **Planes** carry people and things.

plane

## plant

**1.** A **plant** is anything alive that is not a person or an animal. Most **plants** grow in the ground. Flowers, trees, and vegetables are all **plants.**
**2.** Joe puts the pumpkin seeds in the ground. He **plants** seeds every spring. The seeds he **planted** last spring grew into big pumpkins late in the fall.

## plastic

Many things are made from **plastic. Plastic** can be thin or thick. It can be soft or hard. It can be any color. **Plastics** are often used instead of wood, metal, or glass.

## play

**1.** Kelly likes to be in games with her friends. They **play** games after school. Yesterday they **played** baseball for an hour.
**2.** Larry blows into his trumpet. He makes music with it. He **plays** songs that everyone in his neighborhood can hear.

**plant**

**3.** A **play** is a kind of story. People act in **plays** while other people watch them. **Plays** are seen in theaters.

## please

Jack asked his mother for an apple. "Ask in a nice way," she said. "Will you **please** get me an apple?" asked Jack.

## plus

If you add two and four, you get six. Two **plus** four is six. The symbol for **plus** is written like this: +. Two **plus** four is written 2 + 4.

## pocket

A **pocket** is a small bag of cloth. **Pockets** are sewn onto jackets, coats, shirts, and pants. Sandy carries a pen in her **pocket**.

**pocket**

## poem

A **poem** tells a story
About things short or tall.
Or it may share a thought
About things big or small.
The words fit together,
But in groups not too long,
That are then often read
Like the words of a song.
Sad **poems** make people cry,
But happy **poems** are fun.
This **poem** may not be either,
But at least this one is done.

## poet

A **poet** is a person who writes poems.
**Poets** are authors.

## point

1. A **point** is the sharp and very narrow end of an object. Pins and needles have **points**.
2. Mark raises his arm toward the moon. He **points** at it. He **pointed** toward the moon to show his baby brother where it was.

## pole

A **pole** is a long piece of wood or metal.
A wood **pole** is part of a rake.
Telephone **poles** are tall and thick.

**pole**

## police

The **police** are a group of people whose job is to protect everybody. The **police** put people in jail if they break laws.

## pond

A **pond** is a large amount of water that is all in one place. Some **ponds** are big enough to swim in. But they are not as big as lakes.

pond

## pony

A **pony** is a kind of horse. **Ponies** are not as big as other horses.

## poor

**Poor** is the opposite of rich. **Poor** people do not have much money.

## popcorn

**Popcorn** is a kind of food. It is made from small pieces of corn. The corn gets big and soft when it is cooked very quickly in hot oil.

pony

## possible

Anything that is **possible** might happen. It is **possible** to teach some parrots to talk.

## post office

A **post office** is a building. When you mail a letter, it goes to a **post office**. From the **post office** your letter goes to the person you sent it to. Most towns have **post offices.**

## pot

A **pot** is a deep, round pan made of metal or glass. Most **pots** have handles. People cook food in **pots.**

## potato

A **potato** is a vegetable. Its brown skin is covered with bumps. **Potatoes** grow underground. They can be cooked in several different ways.

**pot, potato**

## pound

A **pound** is an amount of weight. It is about the same as 450 grams. Three apples weigh about one **pound.** One ton is 2,000 **pounds.**

## pour

**1.** Tina lifts a bottle of milk. She turns it so that the milk will go into a glass. She **pours** the milk with care. She **poured** the milk until the glass was full.
**2.** When a lot of rain falls from the sky, it **pours** outside.

## powder

A **powder** is anything made of tiny pieces. Flour and dust are **powders.**

## power

**1.** People have **power** when they can decide things. Police have the **power** to put someone who breaks the law in jail.

**2.** When something is strong, it has **power.** A bulldozer has the **power** to move large piles of dirt.

**3.** When energy is made to do work, it is called **power.** Windmills use wind **power** to do work.

## practice

**1.** Allison plays a song many times on her violin. She **practices** the song so that she will play it well. Allison **practiced** the song for an hour.

**2.** Jean draws pictures a lot. She knows it takes a lot of **practice** to draw well.

## prepare

Bill gets ready to go to camp. He **prepares** for camp every summer. Bill **prepared** for camp when he packed his clothes in a trunk.

## present[1]

Jessica's parents gave her a gift on her birthday. They gave her a **present** they knew she wanted. Jessica got other **presents** from her friends.

## present[2]

The **present** is a part of time. It is between the past and the future. It is the part that happens now. Today is part of the **present.**

## pretend

On Halloween John puts on a clown costume and great big shoes. He **pretends** to be a clown. He is not really a clown. He **pretended** so well that he made everybody laugh and laugh.

## pretty

A flower is nice to look at. It is **pretty.**

## price

A baseball ticket costs five dollars. The **price** of the ticket is five dollars. Ticket **prices** go up every year.

## prince

A **prince** is the son of a king or a queen. **Princes** in fairy tales often fight dragons.

## princess

A **princess** is the daughter of a king or a queen. Some **princesses** become queens and rule their countries.

## print

**1.** George writes his name on a piece of paper. None of the letters touch each other. George **prints** his name with care. He **printed** it at the top of the paper.
**2.** People use machines to make thousands of newspapers. The machines **print** these newspapers.

## prize

Anne brought the best apple pie to the fair. She won a **prize** for it. The **prize** was a gold cup. Her friends also won **prizes** at the fair.

**prize**

## probably

Dan thinks he will go to the beach. But he is not sure. He will **probably** go if the weather is good.

## problem

A **problem** is something that causes trouble. Larry had a **problem** when he tore his pants in school. There are many kinds of **problems.**

## promise

**1.** Ellen tells her family she will write to them from camp. She **promises** to write. Ellen **promised** to write her family twice each week.
**2.** Roger told Jeff he would meet him at four o'clock. Roger made a **promise** to meet Jeff. Roger does not break his **promises.**

## propeller

A **propeller** is part of a machine. It is made of wood or metal. When **propellers** turn, they make planes and boats move.

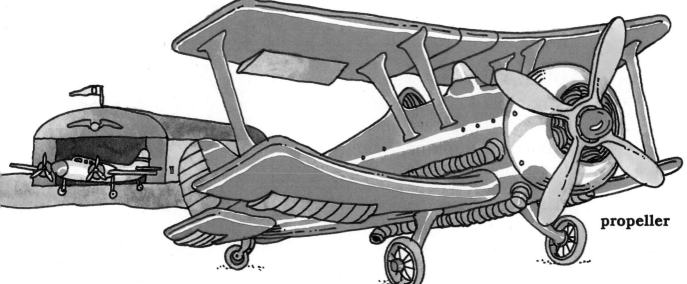

propeller

## protect

Oliver uses an umbrella when it rains. The umbrella keeps him dry. It **protects** him from the rain. The umbrella **protected** him until the wind blew the umbrella away.

## proud

Nancy is glad that she won the race at school.
She practiced a lot for the race.  She is
**proud** that she won it.

## puck

A **puck** is a round, flat object.
It is made of very hard rubber.
**Pucks** are used
in ice hockey games.

## puddle

A **puddle** is an amount
of water that has collected
in one place.  Rain makes
**puddles** on the ground.

## pull

Charlie makes his red wagon
follow him.  He **pulls** the
wagon across the yard.  Last
week he **pulled** it to the store
so he could carry some food
home.

## pumpkin

A **pumpkin** is a large kind
of fruit.   People make funny
faces on **pumpkins**
for Halloween.

## puppet

A **puppet** is a kind of toy.
It looks like a person
or an animal.
Some **puppets** fit
over a hand like
a glove.  Others are
moved by strings
pulled from above.

**puppet**

## puppy

A **puppy** is a young dog. **Puppies** have soft fur and big feet. They like to play with everything.

## push

Charlie puts his hands on his red wagon and makes it go. He **pushes** it across the yard. Yesterday he **pushed** his sister around the yard in the wagon.

## put

Alan picks up his sandwich. He finds a place for it in his lunch box. He **puts** his sandwich in it before he goes to school. Last night he **put** the sandwich in the refrigerator to keep it fresh.

## puzzle

1. A **puzzle** is a game. Some **puzzles** are pieces of paper, metal, or wood that you have to put together to form pictures or objects. Other **puzzles** are problems that you work on with a pencil and paper.
2. A **puzzle** is anything that is hard to understand. It was a **puzzle** to Jane how her sister got to school so quickly.

A B C D E F G H I J K L M
N O P **Q** R S T U V W X Y Z

**Qq Rr**

## quart

A **quart** is an amount of liquid. It is almost as big as a liter. Four **quarts** is one gallon.

## quarter

1. A **quarter** is a kind of coin. Four **quarters** will buy the same things as one dollar.
2. Ann cut the pie into four pieces. Each piece was a **quarter** of the pie.

**quarter**

## queen

1. A **queen** is a woman who rules a country. Most **queens** rule for as long as they live.
2. A **queen** is the name of a card used in some games. This card has a picture of a **queen** on it.

## question

A **question** is a group of words that ask something that you want to know. **Questions** are asked to get answers.

## quick

A deer moves fast. A deer is a **quick** animal.

## quickly

Lightning flashes in a moment. It flashes **quickly.**

## quiet

Meg makes no noise in her house. She is **quiet.**

## quite

1. David has three more pages to read in a book. He is not **quite** done with it.
2. **Quite** means very. It snowed this morning. It was **quite** cold then.

## rabbit

A **rabbit** is an animal. It has long ears.
**Rabbits** hop on their long back legs.
They live in holes in the ground.

rabbit

## race

A **race** is any contest to find which person or
animal is fast. In many **races** people run
from one place to another.

## radio

A **radio** is a kind of machine. It changes energy
sent from other places into sound. Jill and
Cindy listen to music on their **radios.**

## raft

A **raft** is a kind of boat. Some **rafts** are made
of logs. Other **rafts** are boards
on big metal barrels.

raft

**railroad**

### railroad

A **railroad** is the metal path that trains ride on.
Some **railroads** go across rivers on bridges.

### rain

1. **Rain** is water that falls in drops from clouds.
2. When it **rains,** the ground gets wet. Erica was
outside when it **rained** today. She got wet too.

### rainbow

A **rainbow** is like a ribbon of many colors.
Sometimes **rainbows** are across the sky.

### raise

1. Jim takes care of the
vegetables in his garden.
He **raises** them to eat.
Last year he **raised**
lettuce and tomatoes.
2. Jennifer lifts up
her hand.
She **raises** her hand.
3. Fruit costs more
in the winter.
Supermarkets **raise** the price
of fruit then.

**rainbow**

## rake

**1.** A **rake** is a kind of tool. It is made of many pieces of metal or wood joined together like fingers. These pieces are at one end of a long wood pole. **Rakes** are used to gather leaves together.

**2.** Every October Paul works on his lawn. He **rakes** the leaves that fall from the trees. He **raked** them into a neat pile.

## ran

Judy **runs** to school. She **ran** to school because she was late.
**Ran** is a form of **run.**

## rang

A bell **rings** when it is hit. The school bell **rang** when school began.
**Rang** is a form of **ring.**

## rat

A **rat** is an animal. It has four legs and a long tail. **Rats** look like large mice.

rake

rat

## reach

**1.** Rachel puts her hand toward a banana. She **reaches** for the banana. After she ate it, she **reached** for another one.

**2.** It does not snow much near the ocean. Snow becomes rain before it **reaches** the ocean.

## read

Sarah looks at the words in a book.
She knows what they mean.
She **reads** the words.
Yesterday Sarah **read** a story
to her brother.

## ready

Eleanor puts dinner
on the table. The food is cooked
and hot. Dinner is **ready.**

record

## real

Some authors write stories about people who
actually lived. These are stories
about **real** people.

## really

Pine needles do not look like
other leaves. But they are
**really** leaves too.

## reason

Ben was late to school
because it snowed the night
before. That was the **reason**
he was late. Snow and winter
storms are some of the **reasons**
why schools close.

## record

A **record** is a round, flat
object. Phonographs play
music from **records.**

## rectangle

A **rectangle** is a shape.
**Rectangles** have four
sides and four corners.

rectangle

R

## red

**Red** is a color. Fire engines are **red**.

## refrigerator

A **refrigerator** is a machine. It keeps foods and drinks cold. Most people have **refrigerators** in their kitchens.

## remember

Laurie still knows what she did on her last birthday. She **remembers** everything that happened. She also **remembered** that it snowed the whole day.

## repeat

An echo is a sound that is heard several times. It **repeats** itself. The echo in the mountains **repeated** itself five times.

## reptile

A **reptile** is a kind of animal. Snakes, turtles, and alligators are **reptiles.**

reptile

## respect

1. Most people think their laws are good and right. They have **respect** for their laws.
2. A person who **respects** a law will not break it. Laws should be **respected.**

## rest¹

If Stacy is tired, she sleeps. She **rests** while she sleeps. After Stacy **rested,** she was not tired.

## rest²

Nora read half of a book. She will read the **rest** of the book tomorrow.

## restaurant

A **restaurant** is a place where people buy meals. Most **restaurants** make many different foods.

## return

1. The birds come back north every spring. They **return** to the north when the weather gets warm. The birds **returned** early this year.
2. Megan takes the books to the library after she has read them. She **returns** them.

## rhinoceros

A **rhinoceros** is a huge animal. It has short legs, gray skin, and one or two horns.

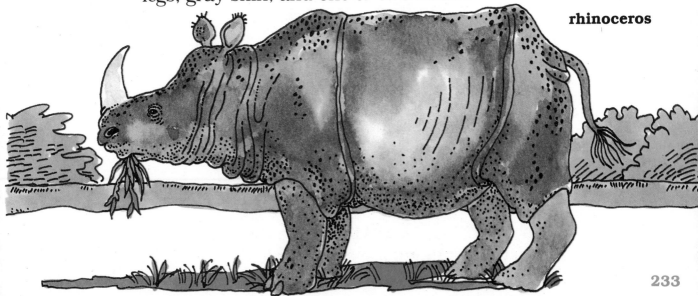

rhinoceros

## rhyme

**1.** Two words that **rhyme** end with the same letters. Cook **rhymes** with book. Group **rhymes** with soup. Tall and small **rhymed** in a poem.
**2.** Reach and beach end with the same letters. They are a **rhyme.** Many poems have **rhymes** in them.

## ribbon

A **ribbon** is a piece of cloth or paper. It is long and narrow. **Ribbons** come in many bright colors.

**ribbon**

## rice

**Rice** is a kind of food. It comes from a kind of grass that grows in warm places. Pieces of **rice** are short and white when they are cooked.

## rich

**Rich** is the opposite of poor. **Rich** people have a lot of money.

## ridden

Tina **rides** a horse across the field. She has **ridden** the horse many times. **Ridden** is a form of **ride.**

## riddle

A **riddle** is a question that has a funny answer. "When does Thursday come before Wednesday?" is one **riddle.** "In a dictionary" is the answer. **Riddles** are a kind of joke.

## ride

**1.** Tina sits on a horse as it moves across a field. She **rides** the horse. She **rode** it for an hour.

**2.** Carol goes to school in a bus. She **rides** on the bus.

**3.** Jean goes to the country in a car. She goes for a **ride** in the country. She has gone for many **rides** in the country.

**4.** A **ride** is a machine. People go on **rides** at fairs. The **rides** turn them around or carry them through the air.

## right

**1.** **Right** is the opposite of wrong. People should be honest. It is **right** for them to be honest.

**2.** **Right** is the opposite of wrong. Nobody should climb on a horse over its head. The **right** way to climb on a horse is from the side.

**3.** **Right** is the opposite of left. People drive cars on the **right** side of the road.

## ring[1]

**1.** A **ring** is a circle with an empty center. The circus tiger jumped through **rings** of fire.

**2.** Jane wears a gold circle on her finger. Her finger fits through the middle of a narrow gold **ring.**

## ring[2]

A bell makes a sound when it is hit. It **rings.** The bell **rang** when school began.

## rise

**1.** The sun goes up in the sky every morning. It **rises** in the east. The sun **rose** later in the morning during the winter.

**2.** Fruit costs more in the winter. Its price **rises** when the fruit must come from far away.

## risen

**1.** The sun **rises** in the east. It had **risen** before Michael woke up.

**2.** The price of fruit has **risen** this winter. **Risen** is a form of **rise.**

## river

A **river** is a wide path of water that has land on both sides. Some **rivers** are hundreds of miles long.

river, road

## road

A **road** is a wide path. It can go through fields, forests, and towns. People travel in cars, trucks, and buses over **roads.**

## roast

Sandy cooks a turkey in the oven. The turkey cooks on all sides at once. The turkey **roasts** in the oven. Sandy **roasted** a turkey for dinner.

## rob

Five people take money from a bank. The money is not theirs. They **rob** the bank. They **robbed** the bank at night.

## robin

A **robin** is a kind
of bird. The front
of a **robin** is red. **Robins**
sing when spring begins.

## robot

A **robot** is a machine.
It is built to do some
of the jobs people
do. Some **robots**
look like people.

robin

## rock

**Rock** is the hard part
of the earth. It is not made of dirt.
Some **rocks** have metal in them.

## rocket

A **rocket** is a machine. Astronauts
travel into space in **rockets.**
Fireworks are small **rockets.**

## rode

Tina **rides** a horse
across the field. She **rode** it
for an hour.
**Rode** is a form of **ride.**

## roll

1. The ball turns over and over again
on the grass. It **rolls** across the grass. The ball
**rolled** toward the trees.
2. A **roll** is a kind of bread. It is round. Many
people eat hamburgers in **rolls.**

robot

## roller skate

A **roller skate** is a skate with four small wheels.
**Roller skates** are used for skating on floors,
sidewalks, or other hard surfaces.

## roof

The **roof** is the top of a building. It covers the building. Some **roofs** are flat. Others are shaped like triangles.

## room

1. A **room** is an area of a building. It usually has four walls. **Rooms** come in many different sizes.
2. Only four people can fit in a small car. There is only **room** for four people in it.

## rooster

A **rooster** is a bird. It is a male chicken that has grown up. **Roosters** make a lot of noise when the sun rises.

## root

A **root** is part of a plant. It grows underground. Plants get food from the ground through their **roots**.

root

## rope

**Rope** is used to tie things. It is made of several pieces of string twisted together. Thick **ropes** are used on boats.

rope

## rose[1]

A **rose** is a flower. It has a good smell. Red **roses** are the ones people know and like best. **Roses** can be pink, yellow, and white too. **Roses** grow on bushes.

## rose²

1. The sun **rises** in the east. It **rose** later in the morning during the winter.
2. The price of fruit **rose** in the winter.
**Rose** is a form of **rise.**

## rough

**Rough** is the opposite of smooth. The bark of an oak tree is **rough.**

## round

**Round** is a shape. Certain objects are **round.** A wheel and a coin are **round.** Most balls are **round.**

## rub

Pat takes a piece of cloth and passes it over the window to get the dirt off. She **rubs** the window with a piece of cloth. Yesterday she **rubbed** her shoes to clean them. Mother **rubs** my back with medicine when my back hurts.

## rubber

1. **Rubber** is something that is easy to push and pull. It is made from the **rubber** plant. It is used for many things. Tires are made of **rubber.** The teacher keeps her papers together with a **rubber** band.
2. When it rains, Larry wears **rubbers** over his shoes to keep his feet dry.

## ruby

A **ruby** is a jewel. It is red. **Rubies** are found in rocks.

## rug

A **rug** is used to cover part of a floor. It is made of wool or other things. Some **rugs** have many colors in them. Some **rugs** come from places far away.

**ruby**

## rule

**1.** A **rule** is a kind of law. The **rules** of basketball explain how basketball should be played. Other games have different **rules.**
**2.** The king makes the laws for the country and decides things. The king **rules** the country. Many kings **ruled** for a long time.

## run

Judy moves her legs quickly. She goes too fast to walk. She **runs.** She **ran** to school because she was late.

## rung

A bell **rings** when it is hit. The school bell had **rung** six times during the day.
**Rung** is a form of **ring.**

## rush

Alice hurries home at noon. She **rushes** home to eat lunch. Alice **rushed** home because she did not have much time to eat.

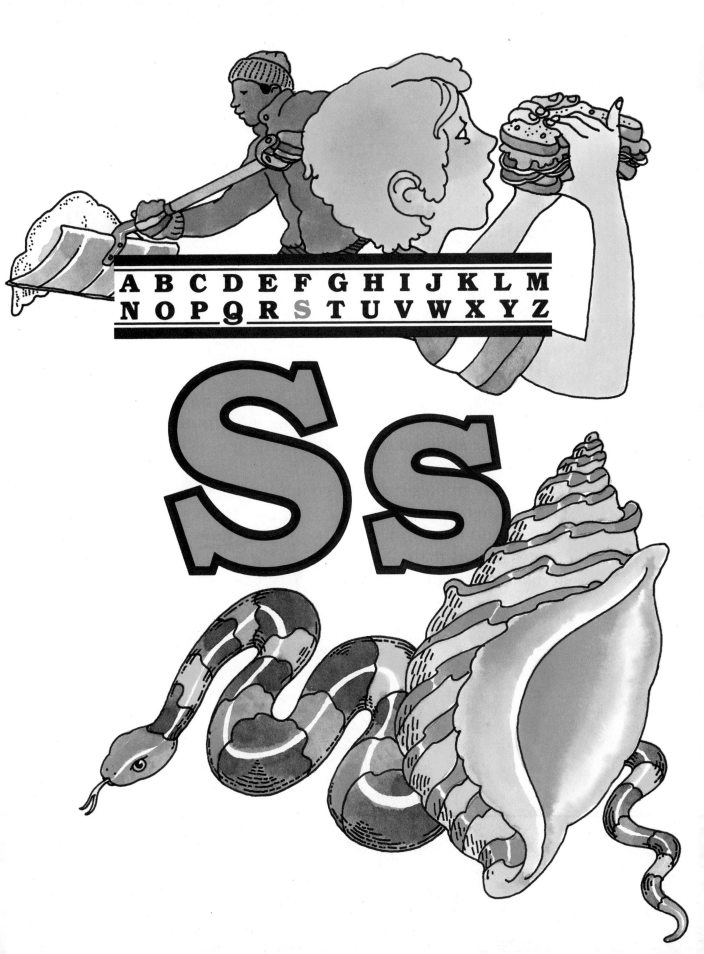

A B C D E F G H I J K L M
N O P Q R S T U V W X Y Z

## sad

Jamie is not happy. He is **sad** because his best friend moved to another town.

## safe

Outside the house there was thunder and lightning and a lot of rain. Inside the house there was no danger from the storm. Everyone inside the house was **safe** from the storm.

## said

When people speak, they **say** words. Eric **said** "Hello!" to his friend on the telephone.
**Said** is a form of **say**.

## sail

**1.** A **sail** is a large piece of strong cloth. **Sails** are tied to tall poles on boats. The wind pushes against the **sails**. It makes the boats go over the water.
**2.** A pirate makes his ship go across the ocean. He **sails** his ship from island to island. Pirates **sailed** their ships across the ocean.

sail

## salt

**Salt** is something people put on food. It looks like white grains of sand. Ocean water has a lot of **salt** in it.

## same

**Same** is the opposite of different. Cathy and Laura both have blue eyes. Their eyes are the **same** color.

## sand

**Sand** is made of tiny grains of rock. It covers beaches at the edge of lakes or oceans. **Sand** can feel rough or soft.

**sand**

## sandwich

A **sandwich** is a kind of food. Meat, cheese, peanut butter, or other foods are put between two pieces of bread to make a **sandwich.** Many people eat **sandwiches** for lunch.

## sang

Elizabeth **sings** the words to a song. She **sang** songs all morning. **Sang** is a form of **sing.**

**sandwich**

S

## sat

Peter **sits** on a chair. He **sat** on it while he read a book.
**Sat** is a form of **sit.**

## Saturday

**Saturday** is the last day of the week. **Saturdays** come after Fridays and before Sundays.

## saucer

Laura pours milk into a **saucer** for the cat. The saucer is a small, shallow dish. **Saucers** usually fit under cups.

## save

**1.** Every week Lucy puts a quarter in a bottle. She **saves** a quarter every week. She **saved** enough quarters to buy a kite.
**2.** Firefighters climb ladders to get people out of houses that burn. They **save** the people inside the houses.

## saw[1]

A **saw** is a tool. It is made of a flat piece of metal with a handle at one end. The edge of the metal piece looks like alligator teeth. **Saws** are used to cut wood, metal, plastic, and other things.

## saw[2]

People **see** everything with their eyes. Many people **saw** the beautiful sunset yesterday.
**Saw** is a form of **see.**

**saw**[1]

**say**

When people speak, they **say** words.
Eric **said** "Hello!" to his friend
on the telephone.

**scare**

Emily sees monsters
in her neighborhood
on Halloween. She is afraid
of them. They **scare**
her. One of them
**scared** her
when it made
a strange noise.

**scarecrow**

A **scarecrow** is
supposed to look
like a person.
It is made
of a wood pole
covered with old clothes
and a hat. **Scarecrows** are put in corn fields
to scare birds away.

**scarecrow**

**school**

**School** is a place where teachers teach students.
**Schools** are usually open from September until
June. Some stay open during the summer too.

**schoolhouse**

George goes to school in a building. That
building is a **schoolhouse**. Many **schoolhouses**
are red.

**science**

**Science** is what people know about the world
and the life that is in it. The land, the ocean, the
weather, the stars, medicine, plants, and
animals are all parts of **science.**

## scissors

**Scissors** are a kind of tool. Two pieces of metal are joined to make one pair of **scissors.** Each piece has a sharp edge and a hole for the fingers at one end. **Scissors** are used to cut paper or cloth.

scissors

## scratch

**1.** The bear makes thin marks on the tree with its claws. It **scratches** the tree. The bear **scratched** the tree to find insects to eat.

**2.** Laurie got a thin, red mark on one hand while she picked berries. It was a **scratch** from the bush. She got several **scratches** from the bushes before she filled her pail.

## sea

**Sea** is another name for **ocean.** Hundreds of years ago pirates sailed over the **seas.**

## seal

A **seal** is an animal. It has smooth and thick fur. It lives in the ocean. **Seals** are not fish.

seal

## sea lion

A **sea lion** is a kind of seal. **Sea lions** are usually larger than seals. They also have smooth fur.

## season

A **season** is part of the year. It is three months long. The four **seasons** in every year are winter, spring, summer, and fall.

## seat

A **seat** is something to sit on. **Seats** can be covered with leather, cloth, plastic, or other things.

## second¹

A **second** is a very short amount of time. Lightning flashes in about a **second.** One minute is 60 **seconds.**

## second²

The **second** letter of the alphabet follows the first letter. The **second** letter is B.

## secret

Eliza and Kathy built a house in a tree. Only they know about it. The house is a **secret** between them. Eliza and Kathy share many **secrets.**

## see

People **see** everything with their eyes. Many people **saw** the beautiful sunset yesterday.

## seed

A **seed** is part of a plant. New plants grow from **seeds.**

seed

## seem

It looks like the magician takes coins
from behind my ear. But it only **seems** that way.
The coins **seemed** to come from behind my ear.
Actually they were in his hand all the time.

## seen

People **see** everything with their eyes. Many
people have **seen** beautiful sunsets.
**Seen** is a form of **see**.

## seesaw

Jackie sits on one end of the **seesaw** and
George sits on the other end. One end of the
**seesaw** goes up while the other goes down.
**Seesaws** are found in parks for children
to play on.

**seesaw**

## sell

Charlie has a bottle of juice and some paper cups
outside his house. When somebody gives him a
dime, he gives back a glass of juice. Charlie **sells**
a glass of juice for a dime. He **sold** eight glasses
in the morning.

## send

**1.** Frank's mother tells him to go to the store.
She **sends** him there to buy a bottle of milk. She
**sent** Frank to the store because she was too
busy to go.
**2.** Susan mails letters to her family from camp.
She **sends** a letter every week.

## sent

1. Frank's mother **sends** him to the store. She **sent** him there because she was too busy to go.
2. Susan **sent** letters to her family every week. **Sent** is a form of **send.**

## sentence

A **sentence** is usually a group of words. It can also be only one word. This is a **sentence.** Questions are also **sentences.**

## September

**September** is a month of the year. It has 30 days. **September** comes after August and before October.

## set

1. When the sun goes below the edge of the sky, the sun **sets.** Evening began after the sun **set.**
2. A **set** is a group of things that belong together. My mother has two **sets** of blue dishes.

## seven

**Seven** is a number. It is written **7.** 6 + 1 is **7.**

## several

Leslie has more than three friends. She has **several** friends.

## sew

Betsy puts a needle and thread through two pieces of cloth. She **sews** the two pieces of cloth together. Betsy **sewed** the cloth to make a dress.

**sew**

## sewn

Betsy **sews** many things.  She has **sewn**
a cover for her fur coat.
**Sewn** is a form of **sew.**

## shadow

A **shadow** is a dark area with light
around it.  When Peter walks
toward a lamp, his **shadow** is
behind him.  Many **shadows** are seen
during the day when the sun shines.

## shake

**1.** Brad mixes the orange juice.  He makes
the jar go up and down very fast.  He
**shakes** the jar until the juice is mixed.
Yesterday Ellen **shook** the
jar to mix the orange juice.
**2.** Amy moves her head
from side to side.  She **shakes** her head
to say no.

## shaken

**1.** Brad **shakes** the jar to mix the juice.
When he has **shaken** it for a minute,
the juice will be mixed.
**2.** Amy does not want any juice.  She
has **shaken** her head to say no.
**Shaken** is a form of **shake.**

## shape

**1.** Basketballs are round.
Basketballs have a round **shape.**
Footballs and basketballs have
different **shapes.**
**2.** David's uncle cuts a piece of wood
with a knife.  He **shapes** it
into a whistle.  He **shaped** whistles
for David.

**shadow**

## share

Everyone in Rick's class has a piece
of his birthday cake. Rick **shares** the cake
with everyone. After he **shared** it, the cake
was all gone.

## shark

A **shark** is a large fish. One part of it sticks out
of its back like a triangle. **Sharks** have
sharp teeth.

shark

## sharp

Something is **sharp** if it has a point or a thin
edge that can cut. Knives have **sharp** edges.

## she

Sharon is a girl. **She** is a female person.

### sheep

A **sheep** is an animal. It has four legs and is covered with wool. **Sheep** are raised for their wool and their meat.

### shell

A **shell** is the outside part of some animals and plants. It is like a hard skin. Lobsters, turtles, eggs, and nuts all have **shells.**

**sheep**

**shell**
Of a lobster

**shell**
Of a turtle

### shine

Bright light comes from the sun. The sun **shines** with light. It **shone** all day yesterday.

### ship

A **ship** is a big boat. It can go in deep water. **Ships** carry people across the ocean.

**ship**

## shirt

A **shirt** is a kind of clothes. People wear **shirts** on the top half of their bodies. Many **shirts** have buttons on them.

## shoe

A **shoe** is something that covers a foot. It may be made of leather, plastic, or cloth. People wear many different kinds of **shoes.**

## shoelace

Gregg stopped running to tie a **shoelace** on his sneaker. Many shoes have **shoelaces.**

## shone

The sun shines with light. It **shone** all day yesterday. **Shone** is a form of **shine.**

## shook

**1.** Brad **shakes** the jar to mix the orange juice. He **shook** it for a minute.
**2.** Amy did not want any juice. She **shook** her head to say no. **Shook** is a form of **shake.**

## shoot

Fireworks go up into the sky. They **shoot** up from the ground. Firefighters **shot** several fireworks up on a summer holiday.

shirt

shoe

253

## shop

**1.** We get our food from a supermarket. We **shop** there for almost everything we eat. We **shopped** at the supermarket every week.

**2.** A **shop** is a kind of small store. Many **shops** stay open Saturdays.

## shore

A **shore** is the land at the edge of water. A beach is part of a **shore.** Oceans, lakes, ponds, and rivers all have **shores.**

shore

shop

## short

**1. Short** is the opposite of tall. Ponies are **short** horses.

**2. Short** is the opposite of long. The barber cut Jim's hair. Jim's hair was very long. Now Jim's hair is **short.**

**3. Short** means small. Meg is not late for school any more. Meg gets ready for school in five minutes. She dresses in a **short** time.

## shot[1]

Fireworks **shoot** up from the ground.
Firefighters **shot** several fireworks up
on a summer holiday.
**Shot** is a form of **shoot.**

## shot[2]

Alex hit the puck with his hockey stick. His
**shot** went into the net. Alex made three **shots**
that went into the net.

## should

School classes start at nine o'clock. Penny is
supposed to be there then. She **should** be
at school when her class starts.

## shoulder

A **shoulder** is part of the body. **Shoulders**
are between the neck and
the arms. Arms are
joined to them.

shoulder

## shout

Ken talks in a loud voice. He **shouts**
across the room. He **shouted**
"Hello!" to his friend.

## shovel

A **shovel** is a tool. It is made of
a wide piece of metal joined
to a long wood handle. People
dig holes
with
**shovels.**

shovel

## show

**1.** Nancy lets Leslie see her book. Nancy **shows** the book to Leslie. Nancy **showed** Leslie some of her other books too.

**2.** Edward does not know how to skate. Stacey does. She skates for him. She **shows** Edward how to skate.

**3.** A story on television or at the movies is a **show.** Many television **shows** are on each week.

## shut

Jamie opens the door. He walks into the house. He **shuts** the door behind him. The door was closed after Jamie **shut** it.

## shy

Wendy is a quiet person. It is hard for her to talk to other people. Wendy is **shy.**

## sick

Caroline's head and stomach hurt. She feels **sick.** Doctors and nurses take care of **sick** people.

## side

**1.** A piece of paper has a back and a front. The back is one **side** of the paper. The front is the other **side.** Both **sides** are the same size.

**2.** A **side** is an edge. A square has four **sides.**

## sidewalk

A **sidewalk** is a path next to a street. There are many **sidewalks** in the city.

**sidewalk**

## sign

**1.** A **sign** is a symbol. The **signs** for plus and minus are + and −.
**2.** A **sign** is a board or a flat piece of metal with words printed on it. Red street **signs** have STOP on them.

## silence

Where there is no sound, there is **silence.**

## silver

**Silver** is a kind of metal. It is soft and white. It is found underground. **Silver** is used to make coins, jewelry, and other things.

sign

## simple

Alex made a sandwich with ham, cheese, turkey, lettuce, and tomatoes. John's sandwich was not fancy. He made a **simple** sandwich. It only had peanut butter in it.

## since

The circus came to town a year ago. It has not returned yet. The circus has not been here **since** last year.

### sing

Betty makes music with her voice. She **sings** a song. She **sang** songs all morning.

### sister

A girl is a **sister** to the other children in her family. Some families have several **sisters** and brothers.

### sit

Peter rests on a chair. He bends at the middle and at the knees. His weight is on the chair. He **sits** on the chair. Peter **sat** on the chair while he read a book.

### six

**Six** is a number. It is written **6**. 5 + 1 is **6**.

### size

The **size** of something is how big it is. People and clothes come in many different **sizes**.

### skate

1. A **skate** is a kind of shoe. It has a long piece of metal on its bottom. Jessica wears **skates** on the ice.
2. A **skate** is a kind of shoe. It has four wheels on its bottom. Linda wears **skates** on the sidewalk.
3. Alan moves across the ice on **skates**. He **skates** well. Alan **skated** around in circles.

**skate**

## ski

**1.** A **ski** is a long piece of wood, metal, or plastic. Ellen wears **skis** to move quickly over the snow. **2.** Jim moves over the snow on **skis.** He and his friend **ski** down the mountain. They **skied** down the mountain five times.

ski

## skin

**1.** The **skin** is the outside part of the body. It covers almost the whole body. **2.** The **skin** is the outside of many things. Apple **skins** can be red, green, or yellow.

## skunk

A **skunk** is an animal. It is about the size of a cat. **Skunks** have four legs and big tails covered with fur. They have a white stripe on their backs. **Skunks** can make a very strong smell.

skunk

## sky

The **sky** is the air far above the ground. It is blue during the day when no clouds cover it. We see the sun, the moon, and the stars in the **sky.**

S

## sled

A **sled** is a toy. People ride
on **sleds** over snow. **Sleds** are made of
wood, metal, or plastic.

## sleep

People rest at night
for several hours.
They **sleep**
in their beds.
Last night
Angela **slept**
for nine hours.

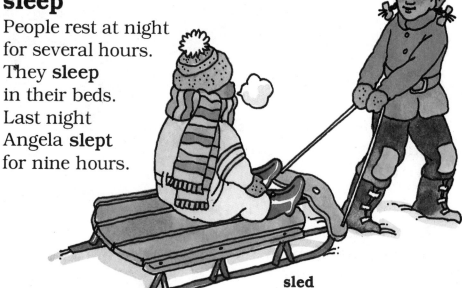

sled

## slept

People **sleep** in their beds. Last night Angela
**slept** for nine hours.
**Slept** is a form of **sleep.**

## slow

**Slow** is the opposite of fast. A lot of time passes
while turtles move. Turtles are **slow** animals.

## small

**Small** is the opposite of big. A butterfly can sit
on a leaf. A butterfly is **small.**

## smell

**1.** Craig knows that the cookies in the oven have
burned. He **smells** the smoke from the oven. He
**smelled** it with his nose.
**2.** Penny holds her nose when she walks
near the swamp. Some swamps **smell** bad.
**3.** The **smell** of a skunk can be very strong.
Skunks and foxes have strong **smells.**

## smile

**1.** People **smile** when they are happy. The ends of their mouths turn up. The crowd **smiled** and laughed at the circus clowns.

**2.** The ends of Erica's mouth are up. Her mouth is shaped in a **smile.** Her **smiles** show that a front tooth has fallen out.

## smoke

**Smoke** is made by many things that burn. It looks like a cloud. **Smoke** can be black, gray, or white.

## smooth

**Smooth** is the opposite of rough. A window has no bumps or points on it. Windows are **smooth.**

## snake

A **snake** is an animal. It has a long, narrow body. **Snakes** do not have arms or legs. They are reptiles. They can be a few inches long or many feet long.

snake

smoke

## sneaker

A **sneaker** is a kind of shoe. It has rubber on the bottom and cloth or leather on the top. People wear **sneakers** when they play sports.

**sneaker**

## sneeze

**1.** When Jill's nose itches, she **sneezes.** The air comes out of her nose and mouth very fast. Jill **sneezed** three times before she stopped.
**2.** Jill **sneezed** a lot. Dennis heard the **sneeze** that Jill made. He heard all three of her **sneezes.**

## snow

**1. Snow** is made of soft white pieces of ice. It comes from the clouds. **Snow** covers the ground like a white blanket.
**2.** When **snow** comes down from the sky, it **snows.** It **snows** every winter in many places. Last year it **snowed** a lot.

## so

**1.** Jamie thought his joke was very funny. He thought it was **so** funny that he laughed for an hour.
**2.** Patrick and Ann liked the same book. Patrick read the book twice. **So** did Ann.
**3.** Chris never carries an umbrella **so** he gets wet when it rains.

## soap

Rachel washes her hands with **soap** after she works in the garden. The **soap** helps take the dirt off her hands. **Soap** makes bubbles when it is mixed with water.

## soccer

**Soccer** is a sport. In **soccer** two teams kick a ball up and down a long field.

## sock

A **sock** is a kind of clothes. People wear **socks** over their feet.

soccer

sock

## soft

1. **Soft** is the opposite of hard. Jane put her head on a feather pillow. It was a **soft** pillow.
2. **Soft** is the opposite of loud. Henry is very quiet in the library. He speaks in a **soft** voice.

## soil

**Soil** is the top part of the ground. Some **soils** are better for plants than others.

## sold

Charlie **sells** glasses of juice outside his house. He **sold** eight glasses in the morning. **Sold** is a form of **sell**.

## some

**1.** Betty has read more than a few of the books in the library. But she has not read most of them. She has read only **some** of the books.
**2.** "Do you want any ketchup on your hamburger?" Carol asked Nancy. "Yes," said Nancy, "I would like **some** ketchup.

## somebody

**Somebody** means some person. We do not know who won the race. But **somebody** must have won it.

## someone

**Someone** is another word for **somebody**. **Someone** turned on the radio in my room while I was outside.

## something

**Something** means some thing. The baby is hungry. She wants some food. She wants **something** to eat.

## sometimes

**Sometimes** means at some times. Water falls from the clouds during the winter. **Sometimes** it rains during the winter and **sometimes** it snows.

## somewhere

**Somewhere** means some place. That robin has a nest in the oak tree. Nobody can see the nest. It is **somewhere** in the branches.

## son

A boy is the **son** of a father and a mother. Some families have several **sons.**

## song

A **song** is music that someone sings. Some **songs** use both instruments and voices together.

## soon

It is almost time for Lauren to go to sleep. She will go to sleep **soon.**

## sorry

Kathy felt sad that Bob was sick. She was **sorry** that he was sick.

## sound

**Sound** is the noise that people and animals hear. Wind, thunder, and fire engines make different **sounds.**

## soup

**Soup** is a kind of food. It is made with water and pieces of meat or vegetables. **Soups** are eaten from bowls.

## sour

Many foods taste **sour.** We put sugar in them to make them sweet. Lemon juice tastes **sour.**

## south

**South** is a direction. **South** is the opposite of north. If you look at where the sun rises, **south** is on your right.

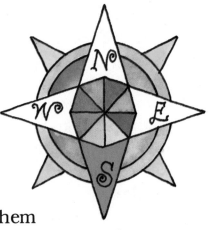

south

## space

1. There is a **space** between letters to keep them apart. There are **spaces** between words too.
2. Rick puts water into a bottle. When the bottle is full, the water fills all of the **space** in it.
3. **Space** is everywhere outside our world. The sun, the moon, and the stars are in **space.** Astronauts have been in **space.**

### speak

People use their voices to talk. They **speak** words. The teacher **spoke** loud enough for everyone in the class to hear.

### special

October 31 is not just any day. October 31 is **special**. It is Halloween.

### spell[1]

Polly puts letters together in the right order to make a word. She can **spell** many words. Polly **spelled** many of the words in this book. She didn't have to look at them first.

### spell[2]

A **spell** is a kind of magic. One witch turned a prince into a frog. It was one of her favorite **spells**.

### spend

Jerry buys a book at the store. He **spends** his money on the book. He **spent** most of the money he had in his pocket.

### spent

Jerry **spends** his money on a book. He **spent** most of the money he had in his pocket.
**Spent** is a form of **spend**.

### spider

A **spider** is a very small animal. It has eight legs. **Spiders** are not insects. However, they catch insects in the webs they make.

**spider**

## spoke

People **speak** words with their voices. The
teacher **spoke** loud enough for everyone
in the class to hear.
**Spoke** is a form of **speak.**

## spoken

People **speak** words with their voices. The
teacher had **spoken** loud enough for everyone
in the class to hear.
**Spoken** is a form of **speak.**

## spoon

A **spoon** is a kind
of tool to eat with. It
looks like a tiny metal
bowl joined to a handle.
People eat soup and
ice cream with **spoons.**

**spoon**

## sport

A **sport** is a kind
of game. People usually get some exercise when
they play **sports.** Baseball, football, hockey,
·tennis, basketball, and soccer are all
different kinds of **sports.**

## spot

1. A **spot** is a small mark that is not the same
color as the area around it. Some animals are
covered with **spots.**
2. A **spot** is a place. Henry pointed to the **spot**
on the street where he found the quarter he had
lost last week.

## spring

**Spring** is a season. It comes after winter and
before summer. Flowers begin to grow
in the **spring.**

### square

A **square** is a kind of rectangle. The sides of a **square** all have the same length. Many floors have **squares** on them.

square

### squirrel

A **squirrel** is a small animal. **Squirrels** have red or gray fur and a big fur tail. **Squirrels** live in trees. They like to eat nuts.

### stable

The farmer keeps his animals in a special building at night. This building is called a **stable.** Horses and cows sleep and sometimes are fed in **stables.**

### stair

A **stair** is one part of a group of steps. Nick walks up the **stairs** in his house. The **stairs** are the way to get from one floor of the house to another.

stair

### stamp

A **stamp** is a small piece of paper. It has a picture, numbers, and words on it. People put **stamps** on letters or packages they send through the mail. They buy the **stamps** to pay for the mail.

stamp

Sara Smith
20 Elm St.
Boston, Ma
02210

## stand

When Leslie **stands** on her feet, her legs are straight. Only her feet touch anything. Leslie **stood** on the sidewalk until she got on the bus.

## star

1. A **star** looks like a tiny dot of light in the sky at night. There are millions and millions of **stars.** Our sun is a **star** too.
2. A **star** is a shape. It has five or six points.
3. A **star** is an important person in a movie, play, or television show. Some shows have several **stars.**

star

## stare

Billy gives a long, curious look at Mr. Baxter's new suit. Billy **stares** at Mr. Baxter's new suit. Yesterday Billy **stared** at the candy in the windows of the supermarket.

## start

1. School begins at nine o'clock. That is when school **starts.** School **started** late on the day of a storm.
2. A car will not go if the engine is off. The engine **starts** when the key is turned. Now the car can move.

## station

A **station** is a special place or building. A fire **station** is a place where fire engines are kept. The police can be found at a police **station.** People buy gas at a gas **station.** Television shows come from television **stations.** You can get on a bus at a bus **station.**

## statue

A **statue** is a kind of art.  **Statues** are made from stone, clay, wood, or metal. Statues can look like people or animals or can have any kind of shape.

statue

## stay

Janet does not go home after school is over.  She **stays** at school to play baseball. Janet **stayed** at school for an extra hour.

## steady

Julie's hands do not shake. She has very **steady** hands. She wants to become a doctor.

## steak

A **steak** is a kind of beef. Many people like to eat **steaks** that are not cooked too much.

## steam

When water gets very hot, it becomes **steam**. **Steam** looks like white smoke.  Water changes into **steam** when it boils.

## step

**1.** The baby lifted her foot and put it down.  She took one **step** forward.  She took three **steps** before she fell down.
**2.** A **step** is a flat place on which people can go up or down.  We walk up the **steps** to the front door of the house.
**3.** Eric walks forward in the parade.  He **steps** forward with the rest of the band.  Eric **stepped** forward while he played his drum.

## stick

**1.** A **stick** is a piece of wood. Many **sticks** are parts of branches. Dogs play with **sticks.**

**2.** Alex plays hockey with a **stick.** It is shaped like an L. He and his friends use their hockey **sticks** to move the puck.

**3.** Jill puts a needle through a shirt. She **sticks** the needle through the cloth. She **stuck** the needle through the cloth many times while she sewed on a button.

**4.** Anne puts a nail into a board with a hammer. The nail doesn't go in all the way. It **sticks** out from the board.

**5.** Rick puts two pieces of paper together with glue. The glue makes the two pieces of paper **stick** together.

## still

My sister was mad at me yesterday. She is mad at me today too. She is **still** mad at me.

## stomach

The **stomach** is part of the body. Any food we eat goes into our **stomachs.**

## stone

A **stone** is a small piece of rock. Smooth **stones** are found on the shores of oceans, lakes, and rivers.

## stood

Leslie **stands** on her feet. She **stood** on the sidewalk until she got on the bus. **Stood** is a form of **stand.**

## stop

The bus moves through the streets. Then it **stops** to let people get off. The bus **stopped** at a station.

## store

A **store** is a place where things are sold. Shoes are sold in shoe **stores**. Furniture is sold in furniture **stores**. A city has many **stores** in it.

## storm

A **storm** is a strong wind with rain or snow. Lightning and thunder come with many **storms**.

## story

A **story** tells about what happened to people and places. Some **stories** are about real things. Some are not.

**store**

## stove

Billy cooks his eggs in the kitchen. He fries them in a pan on the **stove**. **Stoves** also have ovens.

## straight

The edges of a square have no curves or turns. They are **straight** lines.

## strange

Mary felt shy on her first day at school. School was very different from home. She felt **strange** there.

## stranger

Julie just moved to a new town. Nobody in the school knows her. She is a **stranger** there. Every year a few **strangers** come to the school. They are **strangers** only until they meet some new friends.

## straw

A **straw** is a long tube made of paper or plastic. Jan and Laura drink milk through **straws.**

## stream

A **stream** is a narrow path of water. It moves in one direction. **Streams** are not as big as rivers.

## street

A **street** is a road in a city or a town. Large cities have many **streets.**

## string

**String** is used to tie things. It is made from long, strong plants or from a special kind of plastic. Many **strings** are twisted together to make rope. **Strings** come in different sizes and colors like ribbons.

## string bean

A **string bean** is a kind of vegetable. **String beans** are narrow and green. They grow on a bush.

## strip

Lee's belt is a narrow piece of leather. It is a **strip** of leather. **Strips** can have different lengths.

straw

stream

## stripe

A **stripe** is a wide line.
Zebras are covered
with **stripes.**

stripe

## strong

**Strong** is the opposite of weak.
**1.** Oliver can lift two full suitcases
by himself. He has **strong** arms.
**2.** The smell from the skunk went
all over the neighborhood. It was
a **strong** smell.

## stuck

**1.** Jill **sticks** a needle through a shirt.
She **stuck** the needle through the cloth
many times while she sewed on a button.
**2.** The nail did not go in all the way. It **stuck**
out from the board.
**3.** Two pieces of paper had glue between them.
They **stuck** together.
**Stuck** is a form of **stick.**

## student

A **student** is a person who goes to school. One
class has many **students** in it.

## submarine

A **submarine**
is a kind
of ship. It
can travel
underwater.
**Submarines**
are long
and narrow.

submarine

## subtract

When you **subtract** one number from another, the amount of one number is taken away from the amount of the other. Susan **subtracted** 4 from 9. She got 5.

## subway

A **subway** is a train that travels underground through tunnels. **Subways** are used to carry people through large cities.

subway

## such

Kate draws pretty pictures. Tom did not know they were so pretty. He did not know that she drew **such** pretty pictures.

## sudden

The storm arrived very quickly. Nobody expected it. It was a **sudden** storm.

## suddenly

Lightning flashes very fast. It flashes **suddenly.**

## suds

Soap mixed with water makes a lot of bubbles. It makes **suds. Suds** are good for the laundry.

## sugar

**Sugar** is something people put on food. It is made of small white or brown grains that are sweet. Many desserts are made with **sugar.**

## suit

A **suit** is a set of clothes that match. A jacket and pants made from the same cloth is a **suit**. A jacket and skirt made from the same cloth is also a **suit**. Many men and women wear **suits** to their jobs.

## suitcase

A **suitcase** is a kind of box to carry clothes in. It has a handle. **Suitcases** are used when people travel.

suitcase

## summer

**Summer** is a season. It comes after spring and before fall. No other season is as hot as **summer.** During most **summers** a lot of schools are closed.

## sun

The **sun** is a kind of star. It is yellow. We see it shine in the sky during the day. The **sun** gives us heat and light.

## Sunday

**Sunday** is the first day of the week. **Sundays** come after Saturdays and before Mondays.

## sung

Betty **sings** a song. She has **sung** songs all morning.
**Sung** is a form of **sing.**

## sunlight

When the sun shines, there is a lot of light. It is **sunlight. Sunlight** is good for plants.

## sunny

The sun shines on a **sunny** day. There are no clouds in its way.

## sunrise

**Sunrise** is the time when the sun rises. Mornings begin with **sunrises.**

## sunset

**Sunset** is the time when the sun sets. Evenings begin with **sunsets.**

sunset

## supermarket

A **supermarket** is a big store where food is sold. Meat, milk, eggs, vegetables, cereals, and fruits are all sold in **supermarkets. Supermarkets** are usually very full of people.

## supper

**Supper** is a meal. Many people eat **suppers** in the evening.

## suppose

Matthew thinks it will rain tomorrow. He **supposes** it will rain a lot. He **supposed** it would rain yesterday too. It did not. It snowed instead.

supermarket

### sure

Tony knows he has read all of his books. He is **sure** he has read all of them.

### surface

A **surface** is the outside or top part of something. Boats sail on the ocean's **surface.** Roads are covered with different **surfaces.**

### surprise

**1.** We expected today to be sunny. When the rain started, it was a **surprise** to us. Some **surprises** are not as much fun as others.
**2.** Joe's friends have planned a birthday party for him. Joe does not know about it. His friends **surprise** him with the party. They **surprised** him in his house.

### swallow

Jim makes a piece of apple go from his mouth to his stomach. He **swallows** part of the apple. He **swallowed** the whole apple in a few bites.

### swam

Peter **swims** well. He **swam** across the lake. Swam is a form of **swim.**

### swamp

A **swamp** is an area of land. It is soft and wet. Frogs and mosquitoes live in **swamps.**

### swan

A **swan** is a large bird. It has a long neck. **Swans** spend a lot of time on ponds or lakes. They are usually white.

swan

## sweater

A **sweater** is a kind of clothes. People wear **sweaters** over their shirts. Many **sweaters** are made of wool.

## sweet

Many foods have a **sweet** taste. The taste comes from sugar or syrup. Candy, cake, and cookies are **sweet**.

## swim

Peter moves his arms and legs in the water. He **swims** well. He **swam** across the lake.

## swing

**1.** Ned tries to hit a baseball with his bat. He **swings** the bat at the ball. Ned **swung** at the ball and hit it.

**2.** A **swing** is a seat held by ropes or chains to a high bar. Children play on **swings** in parks or in their yards.

sweater

swing

## swum

Peter **swims** well.  He had **swum** across the lake yesterday.

**Swum** is a form of **swim.**

## swung

Ned **swings** at a baseball with his bat.  He **swung** at the ball and hit it.

**Swung** is a form of **swing.**

## symbol

A **symbol** is a mark or a sign that means something.  Letters of the alphabet are **symbols** for sounds.

## syrup

**Syrup** is a thick, sweet liquid.  It is made from sugar or the juice from some plants.

## system

Language is a group of words used together.  It is a **system** of words.  Some **systems** have many parts that work together.

ABCDEFGHIJKLM
NOPQRSTUVWXYZ

Tt

## table

A **table** is a kind of furniture. It has a flat top and four legs. People sit at **tables** to eat.

## tail

A **tail** is a part of some animals' bodies. Animals have **tails** behind their back legs.

tail

## take

**1.** Donna reaches for a sandwich on the plate. She **takes** the sandwich and eats it. She **took** an apple and ate that too.
**2.** Kerry carries books home from the library. She **takes** the books home to read.
**3.** Craig rides on the bus to school. He **takes** the bus to school in the morning.
**4.** Diane uses her camera to make a picture of her brother. She **takes** a picture of him.
**5.** Dan must play basketball a lot to get better at it. It **takes** a lot of practice to play basketball well.

## taken

**1.** Donna **takes** an apple from the table. She has already **taken** a sandwich to eat.
**2.** Kerry has **taken** some books from the library to read.
**3.** Craig has **taken** the bus to school this morning.
**4.** Diane has **taken** a picture of her brother outside.
**5.** It has **taken** Dan a lot of practice to play basketball well.
**Taken** is a form of **take.**

## tale

A **tale** is a kind of story. Some **tales** have really happened. Fairy **tales** often have magic in them.

## talk

When people speak words, they **talk.** Wendy **talked** to Peter about the picnic.

## tall

**1. Tall** is the opposite of short. A **tall** building goes far above the ground.
**2.** Barbara's height is five feet. She is five feet **tall.**

## tame

Animals that make good pets are **tame.** They are not wild. They will not hurt people.

## taste

**1.** Sally put a berry in her mouth. The berry has a sweet flavor. It **tastes** sweet. The other berries she ate **tasted** sweet too.
**2.** Honey has a sweet flavor. It has a sweet **taste.** Lemons have a sour **taste.** These two **tastes** are very different.

## taught

Kim's parents **teach** her how to read. They **taught** her to read at home.
**Taught** is a form of **teach.**

## taxi

A **taxi** is a kind of car that people pay to ride in. People take **taxis** to go from one place to another.

taxi

## teach

Kim learns how to read from her parents. They **teach** her how to read. They **taught** her to read at home.

## teacher

A **teacher** is a person whose job is to teach other people. Most **teachers** do their work in schools.

## team

A **team** is a group of people who work or play together. In many sports two **teams** play against each other. **Teams** are made up of different numbers of people.

## tear[1]

A **tear** is a drop of salt water. When people cry, **tears** come out of their eyes.

## tear[2]

Bob works at the movie theater. He pulls everyone's ticket into two pieces. He **tears** the ticket in half. Bob **tore** the ticket of everyone who passed him.

## teddy bear

A **teddy bear** is a kind of toy. It is very soft and made of something like fur. **Teddy bears** are often children's favorite toys.

## teeth

**Teeth** means more than one **tooth.** People chew food with their **teeth.**

## telephone

A **telephone** sends voices from one place to another. It uses electricity. **Telephones** are made in two pieces joined by a wire.

telephone

## telescope

A **telescope** helps people see things that are far away. It is made of curved glass or mirrors in a long tube. People can look at the stars with **telescopes.**

telescope

## television

People can see pictures
and hear sounds
on a **television.** These
pictures and sounds are parts
of **television** shows.
**Televisions** use electricity.

**television**

## tell

**1.** George talks about his trip
last summer. He **tells** us all about it.
Ellie **told** us about her trip yesterday.
**2.** Bill knows it is morning. He can
**tell** it is morning because the sky
is light.

## temperature

The **temperature** of something
is how hot or cold it is.
Hot things have high
**temperatures.**
Cold things have
low **temperatures.**

## ten

**Ten** is a number.
**Ten** is written **10.**
9 + 1 is **10.**

**temperature**

## tender

**1.** Roses and tulips are not hard or strong.
They are **tender.**
**2.** The steak we ate tonight was easy
to chew. It was **tender.**

## tennis

**Tennis** is a sport. It is played by two or four
people. In **tennis** a ball is hit many times
across a wide net.

## tent

Some people sleep in a **tent** when they camp. A **tent** is a large piece of cloth. **Tents** are held up with poles and ropes.

tent

## terrible

There was a very bad storm yesterday. It was a **terrible** storm.

## test

A **test** is a way to find out what a person knows. There are questions to answer or other things to do on **tests.** Teachers give many **tests** in school.

## than

Dennis likes summer weather. He does not like winter weather. He likes summer weather much better **than** he likes winter weather.

## thank

Alice thinks Jack was nice to give her a birthday present. She **thanks** him for it. Alice **thanked** her other friends for their presents too.

## that

**1.** I can see two cars on the street. One car is near me. This car is blue. The other car is far away. **That** car is red.

**2.** Ellen sees a pretty dress in a store. It is a dress **that** she wants.

## the

**1.** Joan does not want just any dog. She wants to keep **the** dog that followed her home from school.

**2.** All squirrels have big tails. **The** squirrel is an animal with a big tail.

## theater

A **theater** is a place to watch movies or plays. Some **theaters** are very big.

**theater**

## their

Susan reads her book. Katherine reads her book. They both read **their** books.

## theirs

The students all bought some fish for an aquarium. The fish belong to them. The fish are **theirs.**

## them

Nancy feeds bread to some birds. She feeds **them** bread in her yard.

## themselves

Babies are dressed by other people. They cannot dress **themselves.**

## thermometer

A **thermometer** is a kind of tool. It is used to measure and show temperature. **Thermometers** are often used when people are sick.

thermometer

## then

**1.** It is cold now. It was not cold last summer. The weather was warm **then.**

**2.** Paul pulled his sled up the hill. **Then** he rode his sled down the hill.

## there

**1.** "Where should I put the logs?" Sharon asked her father. He pointed to the fireplace. "Put them **there,**" he said.

**2.** Very few people have stood on the moon. **There** are only a few astronauts who have been to the moon.

## these

Julie wears this pair of socks and this pair of shoes. She wears **these** socks and shoes on her feet.

## they

Sarah and Penny ran on the beach. **They** ran into the ocean.

## they're

**They're** is a short way to write **they are.** Leaves are green in the summer. **They're** red and yellow in the fall.

## thick

1. Fred cannot put his arms completely around the oak tree. Oak trees are **thick.**
2. Honey does not pour quickly. It is a **thick** liquid.

## thin

1. A bicycle tire is a **thin** tire. Truck tires are not **thin.**
2. The front and back of a page are very close together. Paper is **thin.**

**thin**

## thing

1. A **thing** is an object, animal, or plant that is not called by its name. Pictures, machines, models, and other **things** are in museums.
2. Patty gave Carol some of her lunch. It was a nice **thing** for Patty to do.

## think

1. Sam uses his mind to decide which jacket to buy. He **thinks** about which jacket he wants. He **thought** about it until he chose one.
2. Oliver believes he can touch the sky if he climbs a ladder. But he **thinks** he will need a very long ladder.

## third

The **third** letter of the alphabet follows the second letter. The **third** letter is C.

## thirsty

Nora's mouth is dry. She wants to drink a glass of water. Nora is **thirsty.**

## this

Mary eats an apple. She likes the apple she eats now. She likes **this** apple.

## those

Lynne watches ducks on a pond. These ducks play in the water. The ducks she watched yesterday did not play. **Those** ducks only sat on the water.

## though

Jack wants to use his sled outside the house. He can't **though.** There is no snow on the ground.

## thought

**1.** Sam **thinks** about which jacket to buy. He **thought** about it until he chose the jacket he wanted.
**2.** Oliver **thought** he could touch the sky if he climbed a long ladder.
**Thought** is a form of **think.**
**3.** A **thought** is something a person thinks about. Maggie had many happy **thoughts** about her trip last fall.

## thousand

A **thousand** is a number. **Thousand** is written 1,000. Ten hundreds are one **thousand.**
**Thousands** of people go to baseball games on Sundays.

## thread

**Thread** is very thin string. People sew clothes with different **threads**.

## three

**Three** is a number. **Three** is written **3**. $2 + 1 = 3$.

## threw

Roger **throws** the ball to Bill. They **threw** the ball to each other for an hour.
**Threw** is a form of **throw**.

## through

1. A bird flew into the house. It flew **through** the open window.
2. Jessica walked from one side of the field to the other. She walked **through** the field.

## throw

Roger makes a ball go through the air. He **throws** the ball to Bill. Bill catches it. They **threw** the ball to each other for an hour.

thread

throw

## thumb

A **thumb** is a part of a hand. **Thumbs** are short, thick fingers. People have one **thumb** on each hand.

## thunder

**Thunder** is a loud noise. It is made when lightning flashes during a storm.

thumb

# Thursday

**Thursday** is a day of the week. **Thursdays** come after Wednesdays and before Fridays.

# ticket

A **ticket** is a small piece of paper. Dan and Anne bought **tickets** for a movie. With the **tickets** they were able to go into the theater.

ticket

# tie

**1.** George makes knots with a piece of rope. He **ties** the rope into knots. He **tied** knots at both ends of the rope.

**2.** Jeff makes a bow with the two ends of the string in his sneaker. He **ties** his sneaker with great care.

**3.** A **tie** is a kind of clothes. It is a long, narrow piece of cloth. People wear **ties** around their necks and in front of their shirts.

## tiger

A **tiger** is a large animal. It looks like a huge cat. **Tigers** have orange fur and black stripes.

## tight

**Tight** is the opposite of loose. The knot in the rope will not come apart. It is a **tight** knot.

## till

David will not get home before six o'clock. He will not get home **till** six o'clock.

tiger

## time

**1. Time** is what clocks measure. Everything happens while **time** passes. **Time** is the past, the present, and the future. Seconds, minutes, hours, days, weeks, and years are all parts of **time.**

**2.** Cathy lives far away from school. She must walk for a long **time** to get there.

**3.** The movie starts at eight o'clock. That is the **time** when the movie starts.

**4.** Abby is late for school. She does not have **time** to eat breakfast.

**5.** The band played a song twice during the parade. The band played it one **time.** Then they played it again. They played it two **times.**

## tin

**Tin** is a kind of metal. It is used to make cans, toys, and other things.

## tiny

**Tiny** means very small. Ants and flies are **tiny** animals.

## tire

A **tire** is a circle of rubber. It covers the outside of a wheel. Cars have four **tires.**

tire

## tired

Karen played basketball all afternoon. She does not feel very strong now. She wants to eat and rest. Karen is **tired.**

## to

**1.** Birds fly toward the south in the fall. They fly **to** warm places for the winter.
**2.** The supermarket opens at nine o'clock. It closes at six o'clock. It is open from nine **to** six.
**3.** Michael paints his yellow walls with white paint. He changes the walls from yellow **to** white.
**4.** I must do a lot of work today. I have a lot of work **to** do.

toe

## today

**Today** is the day that you read this. It is the day after yesterday. It is the day before tomorrow.

## toe

A **toe** is a part of the foot. People have five **toes** on each foot.

## together

**Together** is the opposite of apart. Diane, Nina, and Mary all walked to school with each other. They walked to school **together.**

## told

George **tells** us about his trip. Ellie **told** us about her trip yesterday.
**Told** is a form of **tell.**

## tomato

A **tomato** is a fruit. It is round.
**Tomatoes** are red or green.
Ketchup is made with them.

## tomorrow

**Tomorrow** is the day after today. It is in the future. If today is Sunday, then **tomorrow** is Monday.

## tongue

A **tongue** is a part of the body. It is in the middle of the mouth. People use their **tongues** to help them eat and speak.

## tonight

**Tonight** is the night between today and tomorrow.

## too

**1.** Allison can play two instruments. She plays the violin. She plays the piano **too.**
**2.** It is very hot today. It is **too** hot to work outside.

**tomato**

## took

**1.** Donna **takes** a sandwich and eats it. She **took** an apple and ate that too.
**2.** Kerry **took** some books home from the library.
**3.** Craig **took** the bus to school.
**4.** Diane **took** a picture of her brother.
**5.** It **took** a lot of practice before Dan played basketball well.
**Took** is a form of **take.**

## tool

A **tool** is an object used to help people work. Hammers, rakes, spoons, knives, and shovels are all **tools.**

tool

## tooth

A **tooth** is a part of the body. It is in the mouth. It is hard and white. People chew food with their **teeth.**

## top

**Top** is the opposite of bottom. There is fur on **top** of Barbara's boots. The **tops** of Barbara's boots have fur on them.

### tore

Bob **tears** tickets in half at the movie theater.
He **tore** the ticket of everyone who passed him.
**Tore** is a form of **tear.**

### tornado

A **tornado** is a very strong wind. It twists
through the air shaped like an ice cream cone.
**Tornadoes** can pull trees out of the ground.

### touch

Carrie puts her finger on the oven. She **touches**
the oven. She burned her finger when she
**touched** the hot oven.

### toward

The boat sailed in the direction of the shore. It
sailed **toward** the shore.

### towel

A **towel** is a rectangle
of cloth. Sally uses a big
**towel** to dry herself
after she takes a bath.
**Towels** come
in different sizes.

**towel**

## town

A **town** is a place where people live and work.
**Towns** are not as big as cities.

## toy

A **toy** is an object that children play with. Wood
blocks, wagons, and dolls are **toys.**

## tractor

A **tractor** is a machine. It has big wheels and a
strong engine. Farmers use **tractors** to help
prepare the ground for vegetable gardens.

tractor

## traffic

There are many cars, trucks, and buses
on city streets. There is a lot of **traffic**
on city streets.

## trail

A **trail** is a path through a field or a forest.
People or animals make **trails.**

## trailer

A **trailer** is like a large room on wheels. A
**trailer** can be pulled by a truck to carry things.
**Trailers** are also used by people as houses or
offices.

## train

A **train** is a group of railroad cars. **Trains** carry people, animals, and other things from one place to another.

train

## trap

**1.** Robert digs a big hole in the ground. He covers it with branches. He makes a **trap** for a tiger. He has caught other tigers in his jungle **traps.**

**2.** Robert catches a tiger in the hole. He **traps** the tiger there. He **trapped** the tiger yesterday.

## trash

**Trash** is something that people do not want. Jack puts the **trash** out on the edge of the street every Thursday.

## travel

Leslie goes from her home to the beach. She **travels** to the beach. She **traveled** there many times during the summer.

## treasure

Gold, silver, and jewels are all kinds of
**treasure.** Hundreds of years ago pirates
gathered **treasures** from all over the world.

## tree

A **tree** is a kind of plant. Most of a **tree** is made
of wood. Leaves grow from its branches. Many
**trees** are very tall.

tree

## triangle

A **triangle** is a kind
of shape. It has only three
sides. The sails
on some boats look
like **triangles.**

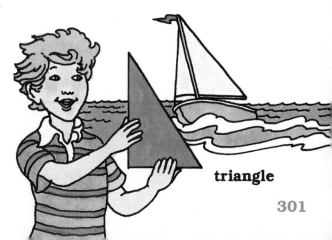

triangle

## trick

**1.** A magician shows a hat to the crowd. Everyone thinks the hat is empty. Then the magician pulls a rabbit out of it. This is a **trick.** Magicians do many **tricks.**

**2.** Eliza's dog does not like medicine. When the dog is sick, Eliza puts the medicine in the dog's food. She **tricks** the dog to make it eat the medicine. She **tricked** the dog to help it get well.

## trick-or-treat

On Halloween children collect candy or fruit from their neighbors at night. They say "Trick-or-treat!" when people open the doors of their houses.

## trip

**1.** Michael rode his bicycle to the lake. He took an hour to get there. The **trip** took an hour. Michael takes a lot of **trips** on his bicycle.

**2.** Kathy hits her foot on a rock and falls. She **trips** over the rock. Kathy **tripped** because she did not see the rock.

## trombone

A **trombone** is an instrument. It is a kind of horn. **Trombones** are made of long metal tubes that fit together. People blow into them to make music.

**trombone**

## troop

A **troop** is a group of people. Many **troops** of school children were at the the town newspaper yesterday to see the machines.

### trouble

**1.** Nick does not add or subtract very well. Problems are hard for him to do. He has **trouble** with numbers. His **troubles** make him unhappy.

**2.** Kelly played in a puddle in her new clothes. Her parents will be angry when they see her. Kelly will be in **trouble** with them.

### trousers

**Trousers** are a kind of clothes. People wear **trousers** over their legs. Many pairs of **trousers** have pockets on the sides near the top.

### truck

A **truck** is a machine. It works like many cars and buses. People use **trucks** to carry animals or objects over roads. **Trucks** can be as small as cars or as big as two buses joined together.

truck

### true

The world is round. This is a fact. It is **true** that the world is round.

## trumpet

A **trumpet** is an instrument. It is a kind of horn. **Trumpets** are made of metal tubes that fit together. They are not as big as trombones. People blow into **trumpets** to make music.

## trunk

**1.** The **trunk** is the thick part of a tree. It grows up the middle of the tree. Branches grow out of **trunks.**

**2.** A **trunk** is a large box. Larry takes a **trunk** filled with clothes to camp.

**3.** A **trunk** is part of an elephant. It looks like a very long nose. An elephant can pick up things with its **trunk.**

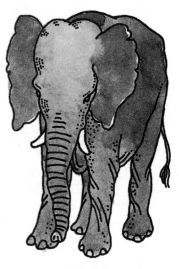

**trumpet**

**trunk**
Of a tree

**trunk**
To keep things in

**trunk**
Of an elephant

## trust

**1.** Florence believes that Sally will do as she says. Florence **trusts** Sally. Florence **trusted** Sally yesterday when Sally said that she would go with her.
**2.** When you believe in people, you have **trust** in them. George has **trust** in his parents.

## truth

Bill always tells things that are true. He always tells the **truth.** He never lies.

## try

**1.** Jan pulls a kite as she runs across the field. She **tries** to get the kite to go up. She ran across the field many times. She **tried** to make the kite go up while the wind blew.
**2.** Lou likes all kinds of ice cream. He will eat any flavor of it. He **tries** every new flavor he finds.

## tub

A **tub** is a round, wide, open kind of bucket. A **tub** is made of wood or metal. **Tubs** are used to store things or to take baths.

## tube

A **tube** is a long, hollow piece of metal, glass, rubber, or plastic. It is round at both ends. Pipes are **tubes.**

## tuck

Larry folds and puts the edges of his shirt into his pants. Larry **tucks** his shirt into his pants. He **tucked** in his pajamas yesterday night too.

## Tuesday

**Tuesday** is a day of the week. **Tuesdays** come after Mondays and before Wednesdays.

## tug

**1.** John pulls hard on his father's coat to get his attention. He **tugs** at his father's coat. He **tugged** at his sister's sleeve yesterday. He also wanted her attention.

**2.** John gave a strong pull at his father's coat. John gave a **tug** at his father's coat. **Tugs** are usually given to get someone's attention.

## tugboat

A **tugboat** is a boat with a strong engine. **Tugboats** push or pull large ships near the shore.

**tugboat**

306

## tulip

A **tulip** is a flower. It is shaped like a cup. **Tulips** come in several colors.

## tunnel

A **tunnel** is a long hole dug underground. Roads or railroads often go through **tunnels**. Some **tunnels** are dug in the ground under lakes or rivers.

## turkey

1. A **turkey** is a bird. It has a long neck. **Turkeys** make strange noises.
2. **Turkey** is a kind of meat. It comes from a **turkey**. It looks like chicken. Many people eat **turkey** on holidays.

tulip

turkey

## turn

1. Bicycle wheels move when the bicycle moves. The wheels **turn**. They **turned** until the bicycle stopped.
2. Craig looks to the left and to the right before he crosses the street. He **turns** his head both ways to look for cars.
3. Wendy looks at one side of a page. Then she **turns** the page to look at the other side.
4. Barbara shared a sandwich with Meg. First Barbara took a bite out of it. Then Meg did. Barbara took another bite. So did Meg. They took **turns** until the sandwich was gone.

## turtle

A **turtle** is an animal. It is a reptile that lives both in the water and on land. **Turtles** can pull their heads, legs, and tails into their shells.

turtle

## twelve

**Twelve** is a number. **Twelve** is written **12**. 11 + 1 is **12**.

## twice

Vicki saw the movie once. Then she saw it again. She saw the movie **twice.**

## twin

A **twin** is one of two children born at the same time to the same parents. Some **twins** look alike.

## twist

**1.** Sally turns two ribbons around each other. She **twists** the ribbons together. Sally **twisted** them together before she put them around a package.
**2.** Carl turned the door key with a **twist** of his hand. It took two **twists** to open both of the locks in the door.

## two

**Two** is a number. **Two** is written **2**. 1 + 1 is **2**.

ABCDEFGHIJKLM
NOPQRSTUVWXYZ

# Uu Vv

## ugly

Monsters are hard to look at.
They are not pretty. They look
mean and terrible. Monsters
are **ugly.**

## umbrella

An **umbrella** is made of cloth
and metal. It opens into the shape
of an upside-down bowl.
People hold **umbrellas**
to keep themselves dry
when it rains.

**umbrella**

## uncle

Any brother of your mother or
your father is your **uncle.** The
husband of any of your aunts is
also your **uncle.** Some people
have several **uncles.**

## under

**1.** A car moves on its wheels. The wheels are
**under** the rest of the car.
**2.** George covers his face with a mask. His face
is **under** the mask.

## underground

**Underground** means under the ground. Rabbits
live in holes that they dig **underground.**

## underline

Tom draws a line under the words that he can't
spell. Tom **underlines** the words. Mary
**underlined** them too.

## understand

Gregory knows what the words in his book
mean. He **understands** the words. He
**understood** all the words that he read.

## understood

Gregory **understands** many words. He **understood** all the words in his book. Understood is a form of **understand.**

## underwater

**Underwater** means under the water. Fish swim **underwater** in the ocean.

underwater

## uneasy

Lily felt nervous about going on the airplane. Lily was **uneasy** about it.

## uneven

The football field was full of bumps. It was not even. It was **uneven.**

## unhappy

**Unhappy** means sad. Diane is **unhappy** because she lost her favorite book.

## unicorn

A **unicorn** is an animal in stories.
It looks like a horse.
All **unicorns** have one horn
that sticks out of their heads.

## uniform

A **uniform** is a special suit
worn by people who belong
to a group. It shows that a
person belongs
to that group. Police,
firefighters, nurses, and
sports teams all wear
**uniforms.**

**unicorn**

## universe

Our world, the sun, the moon, and the stars
are all part of the **universe.** The **universe** is our
name for everything around us.

## unlucky

Every time Beth wants to have a picnic, it rains.
She has bad luck with the weather. Beth is
**unlucky.**

## untie

Tim takes a knot out of a rope. He **unties** the
knot. He **untied** three knots in the rope.

## until

**1.** Alan is in school from nine o'clock to three
o'clock. He is in school **until** three o'clock.
**2.** Leaves are green in the summer. They
change color in the fall. They stay green **until**
the cold weather comes.

## unusual

**Unusual** means not usual. It would be
**unusual** to see an elephant in your garden.

## up

**Up** is the opposite of down.
The fireworks were shot
from the ground to the sky.
They went **up** into the sky.

## upon

Ben sat on a chair. He sat
**upon** the chair.

## upside-down

Eric likes to stand on his head.
When he does this,
everything looks **upside-down**
to him. And he looks
**upside-down**
to other people.

## us

We went to the movies
today. A friend of ours saw **us**
there.

**upside-down**

## use

The carpenter hits the nail with a hammer. She
**uses** the hammer to hit the nail. The carpenter
**used** her hammer and saw to build a house.

## useful

We don't need umbrellas in sunny weather. But
umbrellas are **useful** when it rains.

## usual

Snow falls every winter. It is **usual** for snow to
fall in the winter.

## usually

On most days Tina rides her horse
in the afternoon. She **usually** rides her horse
in the afternoon.

## vacation

A **vacation** is a time when people do not work or go to school. Many children have **vacations** from school every summer.

## valley

A **valley** is the land between mountains or hills. Some **valleys** have rivers at the bottom.

## vanilla

**Vanilla** is a flavor. It is made from a kind of seed. **Vanilla** is used in many desserts.

## variety

David likes to plant different kinds of roses. He plants a **variety** of them in his garden. His parents plant other **varieties** of flowers.

## vegetable

A **vegetable** is a part of a plant that people eat. Lettuce, onions, and potatoes are **vegetables.**

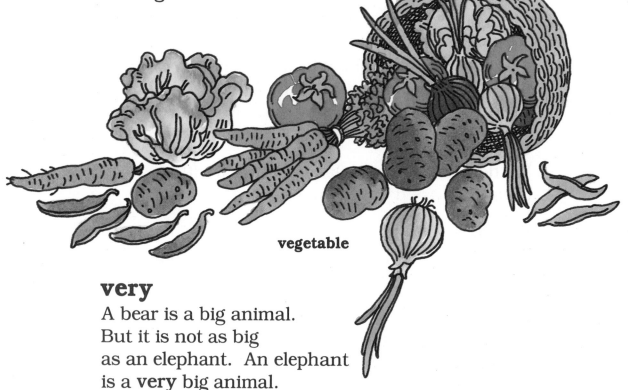

vegetable

## very

A bear is a big animal. But it is not as big as an elephant. An elephant is a **very** big animal.

## village

A **village** is a small group of houses and other buildings. It is often in the country. **Villages** are not as big as towns.

## violin

A **violin** is an instrument. It has four strings. A small stick is pulled across the strings to make music. One orchestra will have several people who play **violins** in it.

violin

## visit

Abby goes to stay with her grandfather in the summer. She **visits** him during her vacation. She **visited** him for a week.

## voice

When Dana talks, she uses her **voice.** People also sing with their **voices.**

## vowel

A **vowel** is a kind of letter. A,E,I,O, and U are **vowels.** Sometimes Y is a **vowel** too.

ABCDEFGHIJKLM
NOPQRSTUVWXYZ

## wagon

A **wagon** is used to carry people or objects. It is made of wood or metal. It has four wheels. Horses or tractors pull large **wagons.** Children pull small ones.

## wait

Eric stands on the sidewalk for ten minutes. He **waits** there for a bus. He **waited** there 20 minutes before a bus came.

## wake

Tracy sleeps at night. She **wakes** up in the morning. After she **wakes** up, she is not asleep. Tracy **woke** up late last Saturday.

## walk

**1.** Pam goes across the room. She puts one foot in front of the other. She **walks** across the room. Pam **walked** across the room and opened a door.
**2.** Henry took a **walk** along the beach. He likes to go on **walks** by himself.

## wall

**1.** A **wall** is a side of a room. It is not the floor or the ceiling. Most rooms have four **walls.**
**2.** A farmer put a **wall** of stones along the edge of a field.

**wall**

## want

Peter would like to go outside. He **wants** to go out after he eats breakfast. He **wanted** to walk in the forest.

## warm

Becky puts a pizza in the oven. The pizza gets **warm** from the heat of the oven. When it gets hot, it will cook.

## was

Jerry's birthday is tomorrow. He will **be** seven years old. He **was** six last year.
**Was** is a form of **be.**

## wash

Carol cleans her hands with soap and water. She **washes** her hands. She **washed** them before she ate dinner.

## wasn't

**Wasn't** is a short way to write **was not.** The wind **wasn't** weak during the last summer hurricane.

## watch

**1.** Meg looks at the rainbow. She **watches** the rainbow in the sky. Meg **watched** the rainbow until it disappeared from the sky.
**2.** A **watch** is a small clock. Many people wear **watches** on their arms.

watch

## water

**Water** is a kind of liquid. It is clear. Oceans, lakes, ponds, rivers, and streams are made of **water**.

## wave

**1.** Jason shakes his hand from side to side. He **waves** his hand at his baby sister. She **waved** back at him.

**2.** A **wave** is part of a lake or an ocean. It rises in the water like a long hill above a field. **Waves** come to the shore and then go back out.

wave

## wax

**Wax** is used to make candles. Bees make one kind of **wax**. Other kinds are made from plants or oil. **Waxes** are soft and easy to melt.

## way

**1.** In the school play Alice must move like a puppet. She walks in a funny **way**. She moves her arms and head in funny **ways** too.

**2.** Jessica can walk over the hill to get to town. She can also walk around the hill. She can go either **way** into town.

## we

My parents and I went to the library. **We** went there together.

## weak

**Weak** is the opposite of strong.
**1.** Nick tries to lift the big box filled with heavy books. But he can't. He is still too **weak** to lift it.
**2.** The skunk left a strong smell near the house three weeks ago. The smell is not strong now. It became **weak.**

## wear

Cathy puts socks on her feet. She always **wears** socks on her feet. She **wore** shoes over her socks.

## weather

Rain, snow, wind, heat, and cold are all kinds of **weather.** Hurricanes, tornadoes, and other storms are the worst kinds of **weather.**

## web

A **web** is a group of threads that a spider makes. Spiders make **webs** in special designs.

## Wednesday

**Wednesday** is a day of the week. **Wednesdays** come after Tuesdays and before Thursdays.

## week

A **week** is an amount of time. One **week** is seven days. There are 52 **weeks** in one year.

## weigh

Michael wants to know how heavy his shoe is. He **weighs** his shoe. The shoe **weighed** one pound.

## weight

The baby weighs 20 pounds. She has a **weight** of 20 pounds. Babies have different **weights** as they grow.

## well[1]

**1.** Allison knows how to play the violin. The music sounds good when she plays it. She plays the violin **well.**

**2.** Edward is sick. But he will feel better soon. When he is **well** again, he will go back to school.

## well[2]

A **well** is a deep hole in the ground. Many **wells** are dug to find water or oil.

## went

**1.** Pam **goes** to school in the morning. She **went** to school late yesterday because the weather was bad.

**2.** Beth is not here. She **went** home.

**3.** The old road **went** around the forest instead of through it.

**4.** Last night Harry **went** to sleep early.

**Went** is a form of **go.**

## were

Alex and Jim like to **be** at the hockey game. They **were** very close to the ice when they **were** there last week.

**Were** is a form of **be.**

## we're

**We're** is a short way to write **we are.** My sisters and I are home for dinner. **We're** very hungry.

## weren't

**Weren't** is a short way to write **were not.** Jim and Mark **weren't** hungry after they ate supper.

## west

**West** is a direction. **West** is the opposite of east. The sun sets in the **west.**

west

## wet

**Wet** is the opposite of dry. Emily went outside in the rain. The rain fell on her. She got **wet.**

## whale

A **whale** is a huge animal that lives in the ocean. It looks like a giant fish, but it is not a fish. No other animals are as big as **whales.**

## wharf

A **wharf** is a kind of large dock. Ships stop there for a while. **Wharves** have many buildings.

## wharves

**Wharves** means more than one **wharf. Wharves** are usually by the sea.

whale

## what

1. Tom said something to Lucy. She did not hear him. "**What** did you say?" she asked.
2. Dennis went to the library. He was not sure about the book he wanted to read. "**What** book should I take out?" he asked.

## whatever

Adam could do anything he wanted on Saturday. He could do **whatever** he thought would be fun.

## wheat

**Wheat** is a kind of plant. **Wheat** seeds are made into flour. Flour is made into bread, cereal, and other things. **Wheat** is an important food.

## wheel

A **wheel** is a round object with flat sides. It is made of wood, metal, rubber, or plastic. **Wheels** turn to help machines move or work. They come in many sizes. A car has four **wheels** covered with tires.

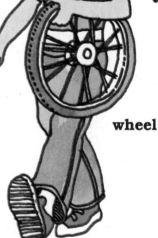

wheel

## when

**1.** George wanted to know what time the basketball game would start. "**When** will the game start?" he asked his sister.
**2.** Wendy cannot change her clothes until she gets home. She can change **when** she gets there.

## where

Lisa could not find her jacket. "**Where** is my jacket?" she asked everyone in her family.

## whether

Our class will have a picnic on Saturday. We will have the picnic even if it rains. We will have it **whether** or not the sun shines.

## which

"Should I wear my blue jacket or my brown one?" Ned asked. "**Which** jacket should I wear?"

## while

Julie buys food at the supermarket during the time it is open. She buys her food **while** the supermarket is open.

## whisper

When Amy speaks in a soft voice, she **whispers**. Amy **whispered** to her friends in the library so she wouldn't bother the other people.

## whistle

**1.** Sam makes a sound like a bird when he blows air out of his mouth. He **whistles** a song. He **whistled** all the songs from a record.

**2.** A **whistle** is a toy. Penny blows into a **whistle** to make a noise like a bird. **Whistles** can be heard from far away.

whistle

## white

**White** is a color. It is the opposite of black. Snow is **white.**

## who

**1.** "Tell me which people were at your party," Andy said to Janice. "**Who** was there?"

**2.** All the people at the football game got wet in the rain. Everyone **who** was there got wet.

## whoever

Nobody knows which person will win the race. **Whoever** wins it will run very fast.

## whole

Everyone in the class was in school today. The **whole** class was there.

## wholly

Barbara did the homework all by herself. She did it **wholly** by herself.

## whom

Yesterday Dan saw Anne in the park. She is the girl **whom** he saw.

## who's

1. **Who's** is a short way to write **who is.**
"**Who's** up in the attic?" Jack asked.
2. **Who's** is a short way to write **who has.**
"**Who's** been outside today?" Sarah asked.

## whose

The teacher saw a glove on the floor. She knew
it belonged to one of her students. "**Whose** glove
is this?" she asked the class.

## why

For what reason did the pirate dig a hole
on the beach? **Why** did he do it?

## wide

1. **Wide** is the opposite of narrow. A river is a
**wide** path of water.
2. Two opposite walls in a room are ten feet
apart. The room is ten feet **wide.**

## width

The room is ten feet wide. Its **width** is ten feet.
Rooms have different **widths.**

## wife

A **wife** is a married woman. She is the **wife**
of the man she married. **Wives** and husbands
are married to each other.

## wild

1. **Wild** is the opposite of tame. **Wild** animals
are not raised by people. They live
by themselves or with one another.
2. **Wild** flowers are flowers that nobody planted.
They grew by themselves.

## will

Gail plans to go to the beach. She **will** go there
tomorrow. She **would** not go yesterday because
it rained.

## win

When two teams play a game, the team that plays better will **win** it. The soccer team **won** most of its games this year.

## wind

**Wind** is air that moves over the ground. Tornadoes and hurricanes have strong **winds** in them.

## windmill

A **windmill** is a machine. It uses the wind to turn large pieces of wood or metal in a circle. These pieces of wood or metal look like propellers. Some **windmills** help make electricity.

## window

A **window** is a hole in a wall. It is usually shaped like a rectangle. **Windows** let light pass from the outside of a building to the inside. They are usually made with glass.

**windmill**

## wing

1. A **wing** is a part of a bird, a bat, and some insects. These animals each have two **wings** to help them fly.
2. A **wing** is a part of a plane. It sticks out like the **wing** of a bird. Most planes have two **wings**.

**wing**
Of a bird

**wing**
Of an airplane

## winter

**Winter** is a season. It comes after fall and before spring. In many places **winters** are very cold.

## wire

A **wire** is a very thin piece of metal. It is shaped like string or thread. It is easy to bend. Electricity moves through **wires.**

wire

## wish

1. Megan wants a dog very much. She **wishes** for a puppy. She **wished** for a puppy every night before she fell asleep.
2. Rachel hopes for something on her birthday. She makes a **wish** when she blows out the candles on her birthday cake. Rachel's birthday **wishes** are a secret.

## witch

A **witch** is a woman in a story who makes magic. There are good and bad **witches** in stories.

## with

1. Bill and Jackie went to the store together. Bill went **with** Jackie to buy some shoes.
2. A giraffe has a long neck. It is an animal **with** a long neck.
3. Brian used a shovel to dig a hole. He dug the hole **with** a shovel.

## without

1. Ned walks to school alone. He walks there **without** anyone else.
2. The sky is blue today. It is a sky **without** any clouds in it.
3. Alan cannot find his gloves. **Without** gloves he won't be able to play in the snow.

## wives

Wives means more than one **wife**. **Wives** and husbands are married to each other.

## wizard

A **wizard** is a kind of magician. In stories **wizards** can do a lot of magic.

## woke

Tracy **wakes** up in the morning. She **woke** up late last Saturday. **Woke** is a form of **wake**.

## wolf

A **wolf** is an animal. It looks like a wild dog. **Wolves** live together in large groups.

## wolves

**Wolves** means more than one **wolf**. **Wolves** live together in groups.

## woman

A girl grows up to be a **woman**. Girls grow up to be **women**.

## women

**Women** means more than one **woman**. Girls grow up to be **women**.

## won

When two teams play a game, the team that plays better will **win** it. The soccer team **won** most of its games this year. **Won** is a form of **win**.

wolf

## wonder

Jill thinks about the color of plants. She **wonders** why so many plants are green. She **wondered** if they were all green for the same reason.

## wonderful

Gary liked the book he just read. He thought it was a great book. He thought it was **wonderful.**

## won't

**Won't** is a short way to write **will not.** Dan **won't** ride his bicycle in the rain.

## wood

1. The trunks and branches of trees are made of **wood. Wood** is used to make furniture, paper, and other things. It is also burned for heat.
2. The **woods** is an area where a lot of trees grow.

**wood**

## wool

**Wool** is a kind of hair. It grows on sheep. **Wool** is used to make clothes and blankets to keep you warm.

## word

A **word** is a sound or a group of sounds. It means something in a language. People speak and write **words** to share their thoughts with other people.

## wore

Cathy **wears** socks on her feet. She **wore** shoes over her socks.
**Wore** is a form of **wear**.

## work

**1.** Bob raked the leaves in his yard. It was hard to do this. It was hard **work** to rake the leaves alone.
**2.** A teacher has a job in a school. Teachers do their **work** in schools.
**3.** Jeff paints his house with a brush. He **works** on the house every day. Jeff **worked** on the house until it was all painted.
**4.** Frank's radio was broken. It didn't **work**. His mother fixed it. Now his radio **works.**

## worker

A **worker** is a person who does some kind of work to earn money. Jerry and Muriel are **workers.**

## world

The **world** is the place where everyone lives. All of the oceans and the land cover the **world.** The **world** is round.

**world**

## worm

A **worm** is a small animal. It looks like a smooth piece of rope. **Worms** live in the ground.

worm

## worn

Cathy **wears** socks on her feet. She has **worn** socks and shoes every day this month.
**Worn** is a form of **wear.**

## worry

Sam cannot sleep. He is not hungry. He is unhappy. His first day of school is tomorrow. He **worries** about what school will be like. He **worried** about it all week.

## worse

**Worse** is the opposite of better. Yesterday there was a lot of rain. The weather was bad. Today there was rain, thunder, lightning, and wind. The weather was **worse** today than it was yesterday.

## worst

**Worst** is the opposite of best. Everyone in Ben's family felt sick. But Ben felt **worse** than anyone else. He felt the **worst** of anyone in his family.

## would

Gail **will** go to the beach tomorrow. She **would** not go yesterday because it rained.
**Would** is a form of **will.**

## wouldn't

**Wouldn't** is a short way to write **would not.** The cat **wouldn't** go into the kitchen because a dog was there.

## wrap

Sharon covers a box with paper. She **wraps** the paper all around the box. Sharon **wrapped** the box for a birthday present.

## wrinkle

A **wrinkle** is a crooked line folded in something. When John frowns, the skin at the top of his nose is covered with **wrinkles**. Jake used an iron to get the **wrinkles** out of his shirt.

## write

**1.** Diane uses a pencil to make a letter on a piece of paper. She **writes** her name on the paper. She **wrote** the date beside her name.
**2.** Authors invent stories. They **write** stories for other people to read.

## written

**1.** Diane **writes** her name on a piece of paper. She has **written** her name many times before.
**2.** Many authors have **written** stories for other people to read.
**Written** is a form of **write**.

## wrong

**Wrong** is the opposite of right.
**1.** People should not rob each other. It is **wrong** for people to rob each other.
**2.** "How many feet are there in one yard?" the teacher asked. "Four," said Margo. "That is the **wrong** answer," said the teacher. "There are three feet in one yard."

## wrote

**1.** Diane **writes** her name at the top of a piece of paper. She **wrote** the date beside her name.
**2.** An author **wrote** a story for other people to read.
**Wrote** is a form of **write**.

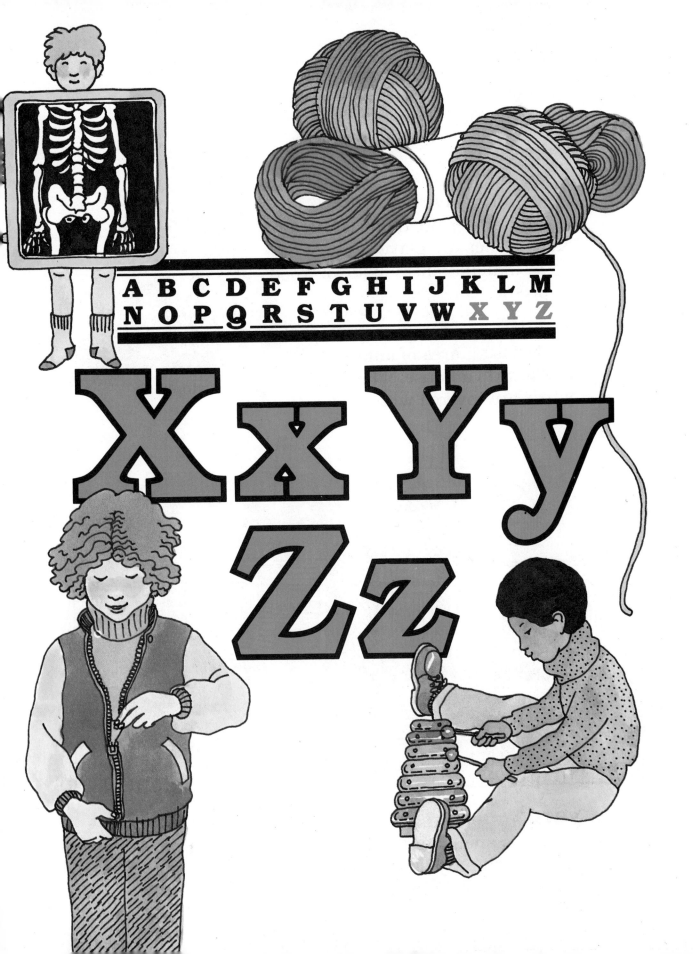

ABCDEFGHIJKLM
NOPQRSTUVWXYZ

Xx Yy
Zz

## x-ray

An **x-ray** is a kind of energy. It can pass through objects. Doctors use **x-rays** to take pictures of the inside of the body.

## xylophone

A **xylophone** is an instrument. It is made of rows of rectangles. The rectangles are different sizes. **Xylophones** are played with small hammers made of wood or plastic.

xylophone

## yard[1]

A **yard** is an area of land. It is next to a house. Many **yards** are covered with grass.

yard[1]

## yard[2]

A **yard** is an amount of length. Three feet are in one **yard**. A **yard** is almost as long as a meter. A football field is one hundred **yards** long.

## yarn

**Yarn** is a kind of loose string. It is made from wool, cotton, or other things. Different **yarns** are used to make sweaters and socks.

## yawn

George is tired. He opens his mouth wide and takes a deep breath. He **yawns** when he is ready to sleep. George put his hand over his mouth as he **yawned.**

**yarn**

## year

A **year** is an amount of time. Each **year** begins in January and ends in December. All **years** have 12 months.

## yell

Jill and Dean are at the hockey game. The game is very exciting. Everyone shouts a lot. Jill and Dean **yell** for their team. They **yelled** the most when their team won.

## yellow

**Yellow** is a color. Bananas are **yellow.**

## yes

**Yes** is the opposite of no. "Will you come to my house after school?" Beth asked Susan. **"Yes,"** said Susan. "I'm glad you will come," said Beth.

## yesterday

**Yesterday** is the day before today. It is
in the past. It snowed **yesterday.** Today the
ground is covered with snow.

## yet

"Has the movie started?" Lynne asked her brother.
"The movie will start soon," he said. "But it has
not started **yet.**"

## you

I wrote this sentence in the past. **You** read this
sentence today.

## young

**Young** is the opposite of old. Babies are very
**young** children.

## your

This book belongs to you. It is not my book.
It is **your** book.

## you're

**You're** is a short way to write **you are. You're**
tall for your age.

## yours

This book belongs to you. It is not mine.
It is **yours.**

## yourself

The oven is very hot. You will burn
**yourself** if you touch it.

## yo-yo

A **yo-yo** is a toy. It has two round pieces
of wood or plastic. There is a string tied
between the round pieces. The **yo-yo** turns
like a wheel, and the string makes it come back.
Some people can do tricks with **yo-yos.**

### zebra

A **zebra** is an animal. It looks like a small horse. **Zebras** have black and white stripes.

### zero

**Zero** is a number. **Zero** is written **0**. $1 + 0$ is still 1. One hundred is written with a one and two **zeros**.

### zipper

A **zipper** is used to hold parts of clothes together. It is made of two metal or plastic sets that fit into each other like teeth. **Zippers** are used on some clothes instead of buttons.

zebra

### zoo

A **zoo** is a place where animals are kept. Many of them are kept in cages. Some **zoos** are like parks.

zipper

# Words that Sound Alike but are Spelled Differently

These words sound alike when we say them aloud. But we spell these words differently. Read the sentences. The sentences tell you the meanings of the words. Study these words. You can build your vocabulary this way.

**ate • eight**

Lee **ate** the big apple.
The number **eight** comes after the number seven.

**bear • bare**

A big **bear** is in the woods.
In winter the oak trees are **bare.** They have no leaves.

**blew • blue**

A strong wind **blew** the tree down.
The paint is **blue.**

**cent • sent**

I found one **cent** in my pocket.
I **sent** the letter today.

**dear • deer**

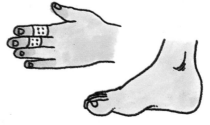

Dear Mary,

People write **dear** to begin letters.
**Deer** are animals that live in the woods.

**heal • heel**

The cuts on my fingers
will **heal**.
The **heel** is the round
back part of the foot.

**hole • whole**

A doughnut has a **hole**
in the middle.
The **whole** class is here.

**knight • night**

The **knight** rides
a horse.
When it is **night,**
we go to sleep.

**meat • meet**

**Meat** is a kind of food.
It can come from
cattle, pigs, turkeys,
and other animals.
My friend and I **meet** after
school on our bicycles.

## pair • pear

I have a **pair** of brown shoes.
Susy ate a **pear** after lunch.

## right • write

Turn **right** at the sign.
We learn to **write** our papers.

## sea • see

The **sea** is another name
for **ocean**.
I **see** better when I wear glasses.

## son • sun

Our neighbors have one **son**.
The **sun** is shining today.

## stair • stare

A **stair** is part of a group of steps.
Billy gives a long **stare** at
the new car.

## tail • tale

The kangaroo has a long **tail**.
A **tale** is a kind of story.
Tales are in books.

# Different Words for
# the Same Thing

There are many words for the same thing. A **cap** is another word for **hat.** A **cap** and a **hat** are the same. Here you can find lots of words like **cap** and **hat.**

**automobile**
**Automobile** is another word for **car.** A car is a machine. It has four wheels, an engine, seats, and windows. Many automobiles are on the roads. John's parents bought a new car.

**band**
Find the words that start with **O.** Put your finger on **orchestra.**

**beautiful**
**Beautiful** is another word for **pretty.** Sarah saw a pretty flower. Rainbows are beautiful.

**big**
Find the words that start with **H.** Put your finger on **huge.**

**boat**
Find the words that start with **S.** Put your finger on **ship.**

**bucket**

**Bucket** is another word for **pail.** A pail or a bucket is used to hold things. Pails and buckets are round at the top and the bottom. They have handles.

**cap**

**Cap** is another word for **hat.** A hat is something you wear on your head. A cap is sometimes smaller than a hat. People who play baseball wear caps. Hats come in many different sizes and shapes.

**car**

Find the words that start with **A.** Put your finger on **automobile.**

**center**

**Center** is another word for **middle.** A white line is in the middle of the highway. It is in the center of the highway.

**cheerful**

**Cheerful** is another word for **happy.** Alan was happy on his birthday. Laura was a cheerful guest at his party.

**coat**

Find the words that start with **J.** Put your finger on **jacket.**

### excellent

**Excellent** is another word for **good.** Helen is a good student. She got an "A" on her report card. It was an excellent report card.

### giant

Find the words that start with **H.** Put your finger on **huge.**

### gift

**Gift** is another word for **present.** Larry gave David a basketball for his birthday. The basketball was a present. David got many gifts on his birthday.

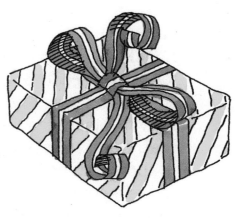

### good

Find the words that start with **E.** Put your finger on **excellent.**

### happy

Find the words that start with **C.** Put your finger on **cheerful.**

### hat

Find the words that start with **C.** Put your finger on **cap.**

### hop

**Hop** is another word for **jump.** A kangaroo is a large animal. It moves very fast when it jumps on its strong back legs. We see the rabbits. They hop across the yard.

**huge**
**Huge, large,** and **giant** are other words for **big.**
Jim made himself a sandwich with turkey, ham, cheese, lettuce, tomato, and three pieces of bread. This was a big sandwich. Elephants are huge animals. A tomato the size of a pumpkin is a giant tomato.

**jacket**
A **jacket** is another word for **coat.** A coat is sometimes long and heavy. A jacket is a short coat. Sarah wears a jacket to the summer fair. She wears a coat in the winter.

**jump**
Find the words that start with **H.** Put your finger on **hop.**

**kettle**
A **kettle** is another word for **pot.** A pot or kettle is a deep, round pan made of metal or glass. Most kettles have handles. People cook food in pots or kettles.

**large**
Find the words that start with **H.** Put your finger on **huge.**

### little

**Little** and **tiny** are other words for **small.** A butterfly is small. A bee is little. An ant is tiny.

### middle

Find the words that start with **C.** Put your finger on **center.**

### ocean

**Ocean** is another word for **sea.** Hundreds of years ago pirates sailed over the sea. The water in the ocean has a lot of salt in it.

**orchestra**

**Orchestra** is another word for **band.** A band is a group of people who play music together. An orchestra is a large group of people who play instruments together.

**pail**

Find the words that start with **B.** Put your finger on **bucket.**

**phone**

Find the words that start with **T.** Put your finger on **telephone.**

**pot**

Find the words that start with **K.** Put your finger on **kettle.**

**present**

Find the words that start with **G.** Put your finger on **gift.**

**pretty**

Find the words that start with **B.** Put your finger on **beautiful.**

**sad**

Find the words that start with **U.** Put your finger on **unhappy.**

**sea**

Find the words that start with **O**. Put your finger on **ocean**.

**ship**

**Ship** is another word for **boat**. Boats carry people and things over the water. Many boats have engines. Some ships have sails.

**small**

Find the words that start with **L**. Put your finger on **little**.

**story**

Find the words that start with **T**. Put your finger on **tale**.

**tale**

**Tale** is another word for **story**. Fairy tales are excellent stories. There are many stories in books.

347

### telephone

**Telephone** is another word for **phone.** A phone sends voices from one place to another. Telephones use a small amount of electricity. A phone rings when someone calls.

### tiny

Find the words that start with **L.** Put your finger on **little.**

### unhappy

**Unhappy** is another word for **sad.** Diane is unhappy because she lost her favorite book. Joe is sad because his pet is sick.